CHILTON'S Repair and Tune-Up Guide

Small Engines

ILLUSTRATED

Prepared by the

Automotive Editorial Department

Chilton Book Company

Chilton Way
Radnor, Pa. 19089
215–687-8200

president and chief executive officer **WILLIAM A. BARBOUR;** executive vice president **RICHARD H. GROVES;** vice president and general manager **WILLIAM D. BYRNE;** associate editorial director **GLEN B. RUH;** managing editor **JOHN H. WEISE, S.A.E.;** assistant managing editor **PETER J. MEYER, S.A.E.;** editor **ROBERT J. BROWN;** technical editor **Philip A. Canal**

CHILTON BOOK COMPANY   RADNOR, PENNSYLVANIA

Library of Congress Cataloging in Publication Data

Chilton Book Company. Automotive Editorial Dept.
   Chilton's repair and tune-up guide: small engines.
   1. Gas and oil engines—Maintenance and repair.
I. Title.   II. Title: Repair and tune-up guide:
small engines.
TJ789.C48 1974   621.43'4   74-779
ISBN 0-8019-5815-6
ISBN 0-8019-5909-8 (pbk.)

CREDITS

BRIGGS & STRATTON CORP.
Milwaukee, Wisconsin

ONAN ENGINE/GENERATOR DIVISION
STUDEBAKER CORP.
Minneapolis, Minn.

KOHLER CO.
Kohler, Wis.

LAWN BOY POWER EQUIPMENT
OUTBOARD MARINE CORP.
Galesburg, Ill.

WISCONSIN ENGINES
TELEDYNE WISCONSIN MOTOR
Milwaukee, Wis.

CLINTON ENGINES CORP.
Maquoketa, Iowa

O & R ENGINES, INC.
Los Angeles, Calif.

TECUMSEH PRODUCTS CO.
Grafton, Wis.

Although information in this guide is based on industry sources
and is as complete as possible at the time of publication, the
possibility exists that the manufacturers made later changes
which could not be included here. While striving for total
accuracy, Chilton Book Company can not assume responsibility
for any errors, changes, or omissions that may occur in the
compilation of this data.

Contents

Introduction
How to Use This Book

The first chapter of this book contains general information pertaining to small, single cylinder engines. The second half of the book is divided into individual chapters for each small engine manufacturer. Each of these sections contains specific information and instructions pertaining to that manufacturer's engines.

When your small engine develops a problem and won't function properly, first turn to the "Small Engine Troubleshooting" section of Chapter One to determine the possible causes of the problem and corrective procedures. If you don't understand how a component operates or what it does, the basic theories of operation of major components are given in the beginning of the chapter. Once you have pinpointed the cause and determined the solution of the problem (and understand why something won't work), then refer to the chapter that deals with your engine to find out exactly how to disassemble, clean, inspect, and reassemble your engine and its various components.

General information such as winterizing, summerizing, general maintenance, and tune-up are also included in the first chapter.

CAUTION: *Whenever working on a recoil starter, exercise* extreme *caution with regard to the recoil spring of the starter. If the spring is allowed to release its tension (unwind) without being checked, it could cause personal injury. You should wear heavy cloth or leather gloves to protect your hands and fingers and some sort of eye or face protection such as goggles or a face mask.*

1 · Engine Operation and General Information

How an Internal Combustion Engine Develops Power

The energy source that runs an internal combustion engine is heat created by the combustion of an air/fuel mixture. The combustion process takes place within a sealed cylinder containing a piston which is able to move up and down in the cylinder. When the flammable gases are ignited, the resulting explosion exerts pressure on the piston, forcing it downward in the cylinder. The piston is connected to a crankshaft by a connecting rod. As the piston is forced downward, it turns the crankshaft, a section of which is off-set from center to provide leverage for the piston and connecting rod. The force causing the downward motion of the piston also causes the crankshaft to turn.

FOUR STROKE ENGINES

The entire series of four events that occur in order for an engine to operate may take place in one revolution of the crankshaft or it may take two revolutions of the crankshaft. The former is termed a two cycle engine and the latter a four cycle engine.

The four events that must occur in order for any internal combustion engine to operate are: intake, compression, expansion or power, and exhaust. When all of these take place in succession, this is considered one cycle.

In a four cycle engine, the intake portion of the cycle takes place when the piston is traveling downward, creating a vacuum within the cylinder. Just as the piston starts to travel downward, a mech-

Intake stroke of a four stroke engine

anically operated valve opens, allowing the fuel/air mixture to be drawn into the cylinder.

As the piston begins to travel upward, the valve closes and the fuel/air mixture becomes trapped in the cylinder. The piston travels upward and compresses the air/fuel mixture. This is the compression part of the cycle.

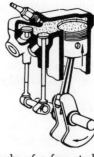

Compression stroke of a four stroke engine

Just as the piston reaches the top of its stroke and starts back down the cylinder, an electric spark ignites the air/fuel mixture and the resulting explosion and rapid expansion of the gases forces the piston downward in the cylinder. This is the expansion or power stroke of the cycle.

Just as the piston reaches the end of its downward travel of the compression stroke,

1

Power stroke of a four stroke engine

The exhaust stroke of a four stroke engine

another mechanically operated valve is opened. The next upward stroke of the piston forces the burned gases out the opened valve. This is the exhaust stroke.

When the piston reaches the top of the cylinder, thus ending the exhaust stroke, the exhaust valve is closed and the intake valve opened. The next downward stroke of the piston is the intake stroke that the whole series of events began with.

TWO STROKE ENGINES

In a two stroke engine, intake, compression, power, and exhaust take place in one downward stroke and one upward stroke of the piston. The spark plug fires every time the piston reaches the top of each stroke, not every other stroke as in a four stroke engine.

The piston in a two cycle engine is used as a sliding valve for the cylinder intake and exhaust ports. The intake and exhaust ports are both open when the piston is at the end of its downward stroke, which is called bottom dead center or BDC. Since the exhaust port is opened to the outside atmospheric pressure, the exhaust gases, which are under a much higher pressure due to combustion, will escape to the outside through the exhaust port. At the same time the exhaust gases are escaping through the exhaust port, a fresh charge of air/fuel mixture is being pumped through the in-

take port, helping to force out the exhaust gases. We say that the air/fuel mixture is being pumped into the cylinder because it is under pressure caused by the downward movement of the piston. The air/fuel mixture is drawn through a one-way valve, known as a reed valve, and into the crankcase by the vacuum caused by upward movement of the piston. When the piston reaches top dead center or TDC and starts back down, the one-way valve in the crankcase is closed by the building pressure caused by the downward movement of the piston. The air/fuel mixture is compressed until the piston moves past the intake port, thus allowing the compressed mixture to enter the cylinder. The piston starts its upward stroke, closing off the intake port and then the exhaust port, thus sealing the cylinder. The air/fuel mixture in the cylinder is compressed and at the same time, a fresh charge of air/fuel mixture is being drawn into the crankcase. When the piston reaches TDC, a spark ignites the compressed air/fuel mixture and the resulting expansion of the gases forces the piston back down the cylinder. The cylinder passes the exhaust port first, allowing the exhaust gases to begin to escape. The piston travels down the cylin-

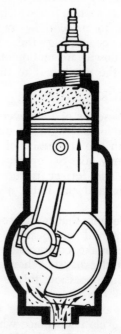

The compression stroke of a two stroke engine; the intake port is open and the air/fuel mixture is entering the crankcase.

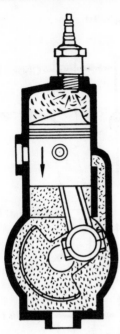

The power stroke of a two stroke engine; the intake port is closed and, as the piston is being forced down by the expanding gases above, the air/fuel mixture in the crankcase is being compressed.

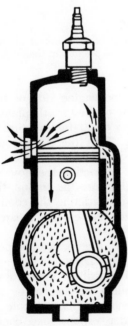

The exhaust stroke of a two stroke engine; the piston travels past the exhaust port, thus opening it, then past the intake port, opening that. As the exhaust gases flow out, the air/fuel mixture flows in due to being under pressure in the crankcase. The next stroke of the piston is the compression stroke and the series of events starts over again.

der a little further and past the intake port. The fresh air/fuel mixture in the crankcase, which was compressed by the downward stroke of the piston, is forced through the intake port and the whole cycle starts over.

Fuel Systems

The fuel system in a small engine consists of a fuel supply or storage vessel, a fuel pump, various fuel lines, and the carburetor. Fuel is stored in the fuel tank and pumped from the tank into the carburetor. Most small engines do not have an actual fuel pump. The carburetor receives gasoline by gravity feed or is drawn into the carburetor by venturi vacuum. The function of the carburetor is to mix the fuel with air in the proper proportions.

PLAIN TUBE CARBURETORS

Carburetors used on engines that run at a constant speed, carrying the same load all of the time, can be relatively simple in design because they are only required to mix fuel and air at a constant ratio. Such is the case with plain tube carburetors.

All carburetors operate on the principle that a gas will flow from a large or wide volume passage through a smaller or narrower volume passage at an increased speed and a decreased pressure over that of the larger passage. This is called the venturi principle. The narrower passage in the carburetor where this acceleration takes place is therefore called the venturi.

A fuel inlet passage is placed at the carburetor venturi. Because of the reduced pressure (vacuum) and the high speed of the air rushing past at that point in the carburetor, the fuel is drawn out of the passage and atomized with the air.

The fuel inlet passage, in most carburetors, can regulate how much fuel is allowed to pass. This is done by a needle valve which consists of a tapered rod (needle) inserted in the opening (seat), partially blocking the flow of fuel out of the opening. The needle is movable in and out of the opening and, since it is tapered at that end, it can regulate the flow of fuel.

The choke of a plain tube carburetor is

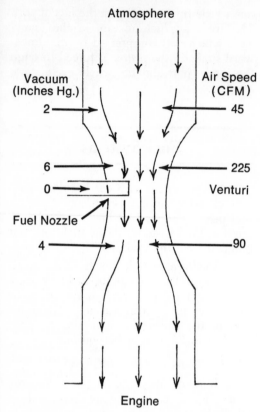

All carburetors operate on the venturi principle. The numbers represent hypothetical ratios of air speed and vacuum in relation to the venturi. Zero vacuum at the fuel nozzle is atmospheric pressure.

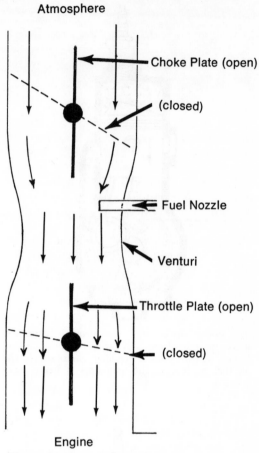

The positioning of the choke and throttle plates in relation to the position of the venturi

located before the venturi or on the atmospheric side. The purpose of the choke is to increase the vacuum within the carburetor during low cranking or starting speeds of the engine. During cranking speeds, there is not enough vacuum present to draw the fuel out through the inlet and into the carburetor to become mixed with the air. When the choke plate closes off the end of the carburetor open to the atmosphere, the vacuum condition in the venturi increases greatly, thus enabling the fuel to be drawn out of the opening.

In most cases the choke is manufactured with a small hole in it so that not all air is blocked off. When the engine starts, the choke is opened slightly to allow more air to pass. The choke usually is not opened all the way until the engine is warmed up and can operate on the leaner fuel/air mixture that comes into the engine when the choke is completely opened and not restricting the air flow at all.

The throttle plate in a carburetor is in-

stalled on the engine side of the venturi. The purpose of the throttle plate is to regulate the flow of air/fuel mixture going into the engine. Thus the throttle plate regulates the speed and power output of an engine. By restricting the flow of air/fuel mixture going into the engine, the throttle plate is also restricting the combustion explosion and the energy created by the explosion.

These are the basic components required for any carburetor to operate on an engine. It is possible for such a carburetor to be installed on an engine and work. However, most engines operate at various speeds, under various load conditions, at different altitudes, and in a variety of temperatures. All of these variations require that the fuel/air mixture can be changed on command. In other words, the simple plain tube carburetor is not sufficient in most cases.

OTHER TYPES OF CARBURETORS

Small engine carburetors are categorized by the way in which the fuel is delivered to the carburetor fuel inlet passage in the venturi (fuel nozzle).

SUCTION CARBURETOR

Next to the plain tube carburetor, this is the simplest in design. With this type of carburetor, the fuel supply is located directly below the carburetor in a fuel tank. In fact, the carburetor and fuel tank are considered one assembly in most cases because a pipe extends from the carburetor venturi down into the fuel tank. When the engine is running, the partial vacuum in the venturi, and the relatively higher atmospheric pressure in the fuel tank, force the fuel up through the fuel pipe and into the carburetor venturi. There is a check ball located in the bottom of the pipe that prevents the fuel in the pipe from draining back into the fuel tank when the engine is shut down. In most engines there is a screen located at the end of the pipe to prevent dirt from entering and blocking the fuel nozzle.

Some suction carburetors are designed with an extra fuel inlet passage for when the engine is idling. This extra passage would be located on the engine side of the throttle plate. Since the vacuum condition in front of an idling engine might not be enough for the fuel to be drawn out at that point, the extra passage is installed behind the throttle plate where the vacuum is great enough to draw the fuel out. This extra fuel inlet passage would also have a regulating needle as does the main inlet.

FLOAT TYPE CARBURETORS

The fuel tank used with float type carburetors is usually located on top of, or at least above, the level of the carburetor. Fuel is fed to the carburetor by gravity. If the fuel tank is located below the level of the carburetor, a fuel pump is employed to pump fuel to the carburetor. Fuel enters the carburetor through a valve and into a float bowl and, as it fills the bowl, a float rises on the surface of the fuel. The float is conected to the inlet valve and, as the level rises, a needle is inserted into the inlet valve and the flow coming into the float bowl is checked. As the fuel is drawn into the main fuel nozzle in the venturi and the level in the bowl drops, the float drops and the needle opens the valve allowing more fuel to run into the bowl.

Although there are many different float carburetors, all operate in this manner.

DIAPHRAGM TYPE CARBURETORS

Fuel is delivered to the diaphragm type carburetor in the same manner as to the float type carburetor.

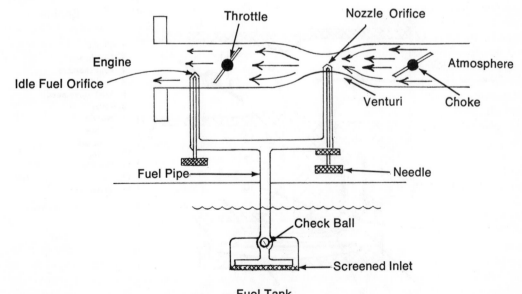

Fuel Tank

Diagram of a suction type carburetor

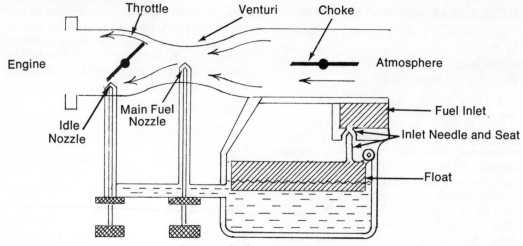

Diagram of a float type carburetor

A flexible diaphragm operates the fuel inlet valve, hence the description "diaphragm carburetor."

Atmospheric pressure is maintained on the under side of the diaphragm by a vent hole in the bottom of the carburetor. The other side of the diaphragm is acted upon by the varying vacuum conditions in the carburetor.

When the vacuum condition in the carburetor is increased by the opening of the throttle plate, and the demand for an increased flow of fuel, the center diaphragm is bellowed upward by the increased vacuum. The center of the diaphragm is connected by a lever to the fuel inlet valve needle. The needle is dropped down away from the inlet valve and fuel is allowed to drop in. When the throttle plate is closed and the need for fuel is reduced, the vacuum condition also decreases and the diaphragm is returned by a spring located on the atmospheric side to its normal flattened position. This closes the fuel inlet valve by raising the needle up into position against its seat.

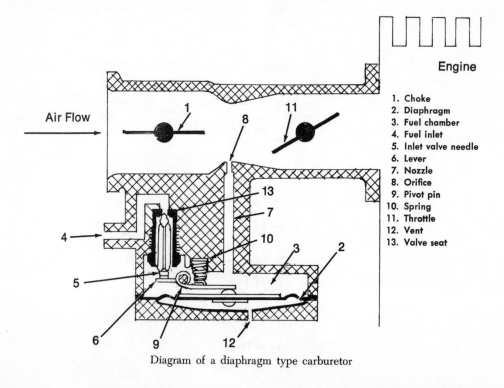

1. Choke
2. Diaphragm
3. Fuel chamber
4. Fuel inlet
5. Inlet valve needle
6. Lever
7. Nozzle
8. Orifice
9. Pivot pin
10. Spring
11. Throttle
12. Vent
13. Valve seat

Diagram of a diaphragm type carburetor

This type of carburetor also has an idle orifice positioned behind the throttle plate to compensate for the low vacuum condition in front of the throttle plate during idle speeds.

Ignition Systems

The function of an ignition system is to provide the electrical spark that ignites the air/fuel mixture in the cylinder at precisely the correct time. The two types of ignition systems used on small engines are either a battery ignition or a magneto ignition. The difference between the two types are where they get their initial electrical charge from. The battery ignition, as the name implies, gets its initial electrical charge from a storage battery. The magneto ignition actually generates its own electricity, thus eliminating the need for a battery as far as ignition is concerned.

Battery and magneto ignition systems are similar in one respect. Both systems take a relatively small amount of electricity, such as 12 volts in the case of a battery ignition, then step up that current to an extremely high voltage, in some systems as high as 20,000 volts.

BASIC BATTERY IGNITION SYSTEM

Battery ignition systems are found mostly on larger one cylinder engines, such as those installed on lawn tractors, larger pumps and generators.

As previously stated, the initial electrical charge comes from a storage battery. The entire ignition system is grounded so that current will flow from the battery throughout the primary circuit.

The ignition system is composed of two segments, the primary circuit and the secondary circuit. Current from the battery flows through the primary circuit and the current that fires the spark plug flows through the secondary circuit. The current in the primary circuit actually creates the secondary current.

When the engine is running, current flows from the battery, through the breaker points, and on to the ignition coil. The current then flows through the primary windings of the coil which are wrapped around a soft iron core and grounded. The primary current flowing through these primary windings causes a magnetic field to be created around the soft iron core of the coil. At precisely the right time, the breaker points open and break the primary circuit, cutting off the flow of current coming from the battery. This causes the magnetic field in the coil to collapse toward the center of the soft iron core. As the field collapses, the lines of force of the magnetic field must pass through the secondary windings of the coil. These windings are also wrapped around the soft iron core of the coil, except that there are many more windings and the wire is thinner

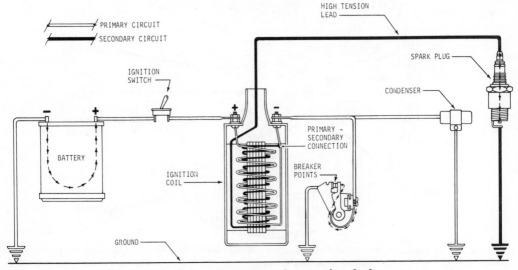

Diagram of a battery type ignition for a single cylinder engine

than those of the primary windings. When the magnetic field collapses and passes through the secondary windings, current flow is induced in the secondary circuit which is grounded at the spark plug. Because the wire of the secondary windings is smaller and there are many more coils around the soft iron core, the current created or induced in the secondary side of the ignition is of much greater voltage than the primary side. This current in the secondary circuit flows toward the grounded end of the circuit which is the spark plug. The current flows down the center of the spark plug to the center electrode. In order to reach the ground, it must jump across a gap to the grounded electrode. When it does, a spark is created and it is this spark that ignites the air/fuel mixture in the cylinder.

MAGNETO TYPE IGNITION SYSTEMS

Magneto ignition systems create the initial primary circuit current, thereby eliminating the need for ignition batteries.

FLYWHEEL TYPE MAGNETOS

On this type of magneto, the flywheel of the engine carries the permanent magnets that are used to create the primary current.

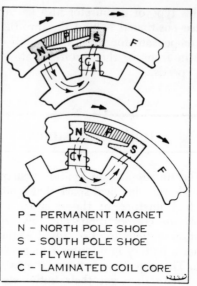

P – PERMANENT MAGNET
N – NORTH POLE SHOE
S – SOUTH POLE SHOE
F – FLYWHEEL
C – LAMINATED COIL CORE

A cutaway view of a flywheel used in a flywheel magneto. The magnets are arranged so that there is a magnetic field covering about ⅓ of the area inclosed by the flywheel. On flywheel magnetos where the coil and core are mounted on the outside of the flywheel, the magnets would be arranged on the outside of the flywheel.

The magnets are arranged so that about ⅓ of the area enclosed by the flywheel is a magnetic field. In the center of the flywheel is a three pronged coil, the three

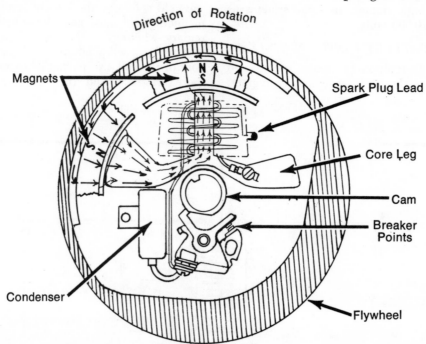

The three legs of the core pass through the magnetic field created by the rotating magnets. In this position the lines of force are concentrated in the left and center core legs and are interlocking the coil windings.

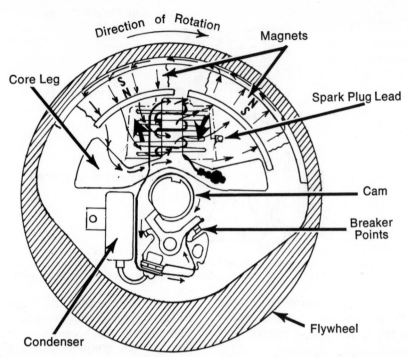

The flywheel has turned to a point where the lines of force of the permanent magnets are being withdrawn from the left and center cores and are being attracted by the center and right cores. Since the center core is both drawing and attracting the lines of force, the lines are cutting up one side and down the other (indicated by the heavy black arrows). The breaker points are now closed and a current is now induced in the primary circuit by the lines of force cutting up and down the center core leg.

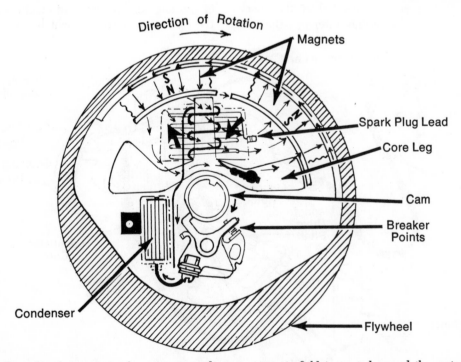

With the induced current in the primary windings, a magnetic field is created around the center core leg (coil). When the field is created around the center leg, the points are opened and the condenser begins to absorb the reverse flow of current.

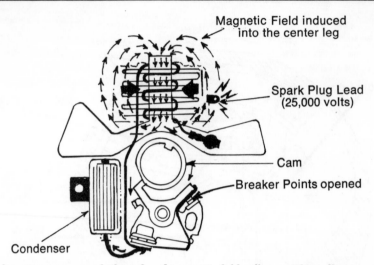

When the breaker points are opened, the induced magnetic field collapses. This collapsing of the magnetic field induces the secondary current into the secondary windings leading to the spark plug.

prongs representing the core. The center prong has the primary and secondary windings and is set up just like a battery ignition coil.

As the magnets in the flywheel pass the two outside prongs, the primary current is induced in the coil and a magnetic field is set up around the center prong. At the point when the primary current is at its strongest which is also the time when the secondary current is needed at the spark plug, the breaker points interrupt the pri-

mary current, causing the magnetic field to collapse through the secondary windings. This induces the secondary current which flows to the grounded spark plug.

UNIT TYPE MAGNETOS

Unit type magnetos operate in the same way as flywheel magnetos.

Permanent magnets are rotated through a mechanical connection to the crankshaft of the engine. The rotating magnets create the primary current which is routed

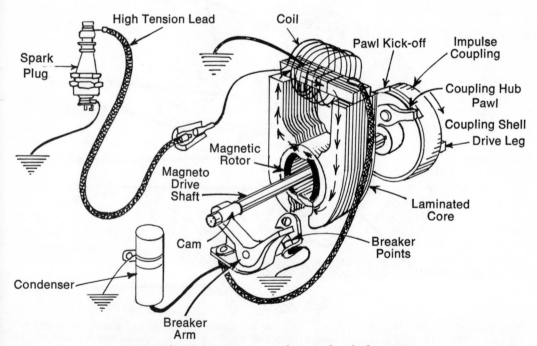

Diagram for a unit type magneto for a single cylinder engine

through the breaker points and on to windings around a soft iron core which is also wrapped by the secondary windings. At precisely the right moment, the points interrupt the primary current, causing the magnetic field to collapse through the secondary windings and inducing the secondary current to the spark plug.

Because the magnets in the unit magneto are driven by the crankshaft of the engine through a gear mechanism, starting is a problem. At starting speeds, the magnets in the magneto cannot be rotated fast enough to create a primary current. To overcome this difficulty, unit magnetos have an impulse coupling on their shaft which drives the magnets. When the engine is turned over at starting speed, a catch engages a coil spring that is wound up in much the same manner as those that propel wind-up toys. As the shaft rotates further, it releases the spring. The spring unwinds rapidly, spinning the magnets fast enough to cause the primary current to be induced in the primary circuit. When the engine starts to run, centrifugal force keeps the catch in the impulse coupling from engaging the wind-up spring.

Lubrication System

FOUR STROKE ENGINES

Most small four stroke engines are lubricated by the splash system. All vital moving parts are splashed with lubricating oil that is stored in the crankcase. The connecting rod bearing gap usually has an arm extending down into the area in which the oil lies. As the crankshaft turns, the arm, commonly called a dipper, splashes oil up onto the cylinder walls, crankshaft bearings and camshaft bearings. In some cases, the connecting rod bearing cap screws are locked in place by lock plates which double as oil dippers. When the lock plate's tabs are bent up beside the cap screws, they extend a little past the tops of the screws and, when they enter the oil, they provide a sufficient splash to lubricate the engine.

In larger four stroke engines, a pressure system is used to lubricate the engine. This type of lubrication system pumps the oil

through special passages in the block and to the components to be lubricated. In some cases, the camshaft and crankshaft have their center drilled so that oil can be pumped through the center and to the bearing journals. Most oil pumps in small engines are gear types.

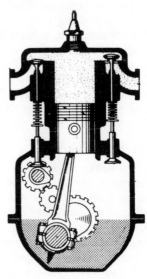

Lubrication of a four stroke engine

In addition to lubricating the engine, oil in four stroke engines has other important functions. One of those additional functions is to help cool the engine. The oil actually absorbs heat from high temperature areas and dissipates it throughout other parts of the engine that are not directly exposed to the very high temperatures of combustion.

Oil in the crankcase also functions as a sealer. It helps to seal off the combustion chamber (top of the piston) from the crankcase, thus maintaining compression which is vital to satisfactory engine operation, while at the same time lubricating the cylinder walls and piston rings.

Engine oil also keeps harmful bits of dust, metal, carbon, or any other material that might be present in the crankcase, in suspension. This keeps potentially harmful abrasives away from vital moving parts.

Most manufacturers recommend a good grade, medium weight detergent oil for their engines. The detergent properties of engine oil do not mean that the oil is capable of cleaning away dirt or sludge deposits already present in the engine. It means that the oil will help fight the formation of such deposits. In other words, the oil keeps the dirt in suspension.

A certain amount of combustion leaks past the piston rings and into the crankcase. Raw gas, carbon, products of combustion, and other undesirable material still manage to make their way into the lubricating oil. The oil becomes diluted by the gas and loses its cooling properties. When it is filled with abrasives, it loses its detergent properties due to chemical reactions, constant heat, and pollutants. The oil can then actually become harmful to the engine. Thus, one can see the need for keeping oil as clean as possible by changing it frequently.

TWO STROKE ENGINES

Since the two stroke engine uses its crankcase to compress the air/fuel mixture so that it can be forced up into the combustion chamber, it cannot also be used as an oil sump. The oil would be splashed around and forced up into the combustion chamber, resulting in the loss of great amounts of lubricating oil while at the same time "contaminating" the air/fuel mixture. In fact the air/fuel mixture would be so filled with oil that it would not ignite.

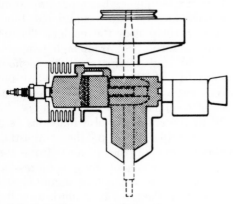

Lubrication of a two stroke engine

A two stroke engine is lubricated by mixing the lubricating oil in with the fuel. A mixture of fuel and lubricating oil is sucked into the crankcase and, while it is being compressed by the downward movement of the piston, enough oil attaches itself to the moving parts of the engine to sufficiently lubricate the engine. The rest of the oil is burned along with the fuel/air mixture. As can well be imagined, the mixture ratio of fuel and oil is critical to a two stroke engine. Too much oil will foul the spark plug and too little oil will not lubricate the engine sufficiently, causing excessive wear and possible engine seizure. Follow the manufacturer's recommendations closely. Most recommend that detergent oils not be used in two stroke engines.

Troubleshooting the Small Engine

FOUR STROKE ENGINES

When troubleshooting the small engine, do not overlook the obvious. Start with the simplest causes first. Check to see if all wires are connected, that the spark plug is in good condition, that the fuel in the tank is clean, and that there is in fact fuel in the tank.

The following troubleshooting guide is meant to be just that, a guide. It lists the most common troubles experienced with two and four stroke single cylinder engines, possible causes of the trouble, and probable remedies. Refer to the specific manufacturer's sections for detailed repair procedures.

PROBLEM: The engine does not start or is hard to start.
CAUSES AND REMEDIES:
1. The fuel tank is empty.
2. The fuel shut-off valve is closed; open it.
3. The fuel line is clogged. Remove the fuel line and clean it. Clean the carburetor if necessary.
4. The fuel tank is not vented properly. Check the fuel tank cap vent to see if it is open.
5. There is water in the fuel supply. Drain the tank, clean the fuel lines and the carburetor, and dry the spark plug. Fill the tank with fresh fuel. Check the fuel supply before pouring it into the engine's fuel tank. Chances are it might be the source of the water.
6. The engine is over choked. Close the throttle and turn the engine over until it starts, then open the throttle. Do not close the choke as far next time.

7. The carburetor is improperly adjusted; adjust it.

8. Magneto wiring is loose or defective. Check the magneto wiring for shorts or grounds and repair it, if necessary.

9. The magneto is faulty. Check the ignition timing and point gap. Replace the magneto if necessary.

10. The spark plug is fouled. Remove, clean, and regap the spark plug.

11. The spark plug is damaged (cracked porcelain, bent electrodes etc.). Replace the spark plug.

12. Compression is poor. The head is loose or the gasket is leaking. Sticking valves or worn piston rings could also be the cause. In any case, the engine will have to be disassembled and the cause of the problem corrected.

PROBLEM: The engine misses under load.
CAUSES AND REMEDIES:

1. The spark plug is fouled. Remove, clean, and regap the spark plug.

2. The spark plug is damaged. Replace the spark plug.

3. The spark plug is improperly gapped. Regap the spark plug to the proper gap.

4. The breaker points are pitted. Replace the points.

5. The breaker point's breaker arm is sluggish. Clean and lubricate it.

6. The condensor is faulty. Replace it.

7. The carburetor is not adjusted properly. Adjust it.

8. The valves are not adjusted properly. Adjust the valve clearance.

9. The valve springs are weak. Replace them.

PROBLEM: The engine knocks.
CAUSES AND REMEDIES:

1. Carbon has built up in the combustion chamber, resulting in retention of excess heat and an increase in compression which causes preignition. Remove the cylinder head and remove the carbon from the head and the top of the piston.

2. The connecting rod is loose or worn. Replace it.

3. The flywheel is loose. Check the flywheel key and keyway in the top of the crankshaft. Replace any worn parts. Tighten the flywheel nut to the specified torque.

4. The cylinder is worn. Rebuild or replace the cylinder.

5. The magneto is not timed correctly. Time the magneto.

6. The engine has overheated. Stop the engine and find the cause of the overheating.

PROBLEM: The engine vibrates excessively.
CAUSES AND REMEDIES:

1. The engine is not mounted securely to the equipment that it operates. Tighten any loose mounting bolts.

2. The equipment that the engine operates is not balanced. Check the equipment.

3. The crankshaft is bent. Replace the crankshaft.

PROBLEM: The engine lacks power.
CAUSES AND REMEDIES:

1. The choke is partially closed. Open the choke.

2. The carburetor is not adjusted correctly. Adjust it.

3. The ignition is not timed correctly. Time the ignition.

4. There is a lack of lubrication or not enough oil in the crankcase. Fill the crankcase to the correct level.

5. The air cleaner is fouled. Clean it.

6. The valves are not sealing. Do a valve job.

PROBLEM: The engine operates erratically, surges, and runs unevenly.
CAUSES AND REMEDIES:

1. The fuel line is clogged. Unclog it.

2. The fuel tank cap vent is clogged. Open the vent hole.

3. There is water in the fuel. Drain the tank, the carburetor, and the fuel lines and refill with fresh gasoline.

4. The fuel pump is faulty. Check the operation of the fuel pump if so equipped.

5. The governor is improperly set or parts are sticking or binding. Set the governor and check for binding parts and correct them.

6. The carburetor is not adjusted properly. Adjust it.

PROBLEM: Engine overheats.
CAUSES AND REMEDIES:

1. The ignition is not timed properly. Time the engine's ignition.

2. The fuel mixture is too lean. Adjust the carburetor.

3. The air intake screen or cooling fins are clogged. Clean away any obstructions.

4. The engine is being operated without the blower housing or shrouds in place. Install the blower housing and shrouds.

5. The engine is operating under an excessive load. Reduce the load and check associated equipment.

6. The oil level is too high. Check the oil level and drain some out if necessary.

7. There is not enough oil in the crankcase. Check the oil level and adjust accordingly.

8. The valve tappet clearance is too close. Adjust the valves to the proper specification.

9. Carbon has built up in the combustion chamber. Remove the cylinder and clean the head and piston of all carbon.

10. An improper amount of oil is mixed with the fuel (two stroke engines only). Drain the fuel tank and fill with correct mixture.

PROBLEM: The crankcase breather is passing oil (four stroke engines only).
CAUSES AND REMEDIES:
1. The engine is being operated at too high rpm. Slow it down by adjusting the governor.

2. The oil fill cap or gasket is missing or damaged. Install a new cap and gasket and tighten it securely.

3. The breather mechanism is damaged. Replace the reed plate assembly.

4. The breather mechanism is dirty. Remove, clean, and replace it.

5. The drain hole in the breather is clogged. Clean the breather assembly and open the hole.

6. The piston ring gaps are aligned. Disassemble the engine and offset the ring gaps 90° from each other.

7. The breather is loose or the gaskets are leaking. Tighten the breather to the crankcase.

8. The rings are not seated properly or they are worn. Install new rings.

PROBLEM: The engine backfires.
CAUSES AND REMEDIES:
1. The carburetor is adjusted so the air/fuel mixture is too lean. Adjust the carburetor.

2. The ignition is not timed correctly. Time the engine.

3. The valves are sticking. Do a valve job.

Operating Precautions, General Maintenance and Preparation for Storage

HOT AND COLD WEATHER OPERATION

There are a few things that must be done when a small engine is to be operated in hot or cold weather, that is, when temperatures are either below 30° F, or above 75° F.

In hot weather, do the following:

1. Keep the cooling fins of the engine block clean and free of all obstructions. Remove all dirt, built up oil and grease, flaking paint and grass.

2. Air should be able to flow to and from the engine with no obstructions. Keep all faring and cover openings free from obstructions.

3. During hot weather service, heavier weight oil should be used in the crankcase. Follow the manufacturer's recommendations as to the heaviest weight oil allowed in the crankcase.

4. Check the oil level each time the fuel tank is filled. An engine will use more oil in extremely hot weather.

5. Check the battery water level more frequently since, in hot weather, the water in the battery will evaporate more quickly.

6. Be on the lookout for vapor lock, which occurs within the carburetor.

7. Use regular grade gasoline rather than premium.

8. Use unleaded gasoline if possible.

9. The most important thing to remember is to keep the engine as clean as possible. Blow it off with compressed air or wash it as often as possible.

CAUTION: *Only wash the engine after it has had sufficient time to cool down to ambient temperatures. Avoid getting water in or even near the carburetor intake opening.*

Precautions for operating a small engine in cold weather are as follows:

1. A lightweight oil should be installed in the crankcase when operating in cold weather. Consult the manufacturer's recommendations.

2. If the engine is filled with summer weight oil, the engine should be moved to a warm (above 60° F) location and allowed to reach ambient temperature before starting. This is because a heavy summer weight oil will be even thicker at cold temperatures. So thick, in fact, that it will be unable to sufficiently lubricate the engine when it is first started and running. Damage could occur due to lack of lubrication.

3. Change the oil only after the engine has been operated long enough for operating temperatures to have been reached. Change the oil while the engine is still hot.

4. Use fresh gasoline. Fill the gas tank daily to prevent the formation of condensation in the tank and fuel lines.

5. Keep the battery in a fully charged condition, since cold weather infringes upon a battery's maximum current output capabilities.

6. Have the battery charged every so often to ensure maximum power output when it is needed most.

STORAGE

If an engine is to be out of service for more than 30 days, the following steps should be performed:

1. Run the engine for 5 to 10 minutes until it is thoroughly warmed up to normal operating temperatures.

2. Turn off the fuel supply while the engine is still running, and continue running it until the engine stops from lack of fuel. This procedure removes all fuel from the carburetor.

3. Drain the oil from the crankcase while the engine is still warm.

4. Fill the crankcase with clean oil and tag the engine to indicate what weight oil was installed.

5. Remove the spark plug and squirt about an ounce of oil into the cylinder. Turn the engine over a few times to coat the cylinder walls, the top of the piston, and the head with a protective coating of oil. Reinstall the spark plug and tighten it to the proper torque.

6. Clean or replace the air cleaner.

Refer to the manufacturer's recommendations.

7. Clean the governor linkage, making sure that it is in good working order and oiling all joints.

8. Plug the exhaust outlet and the fuel inlet openings. Use clean, lintless rags.

9. Remove the battery and store it in a cool place where there is no danger of freezing. Do not store any wet cell battery directly in contact with the ground or cement floor, as it will establish a ground and discharge itself. A completely discharged battery will never be able to be brought back to its original output capacity. Store the battery on a work bench or on blocks of wood on the floor.

10. Wipe off or wash the engine. Wash only after the engine has had time to cool down to ambient temperature and avoid getting water in the carburetor intake port.

11. Coat all parts that might rust with a light coating of oil. Paint all non-operating parts with a rust inhibiting paint.

12. Provide the entire unit with a suitable covering. Plastic is good where the application and removal of sunlight will not promote the formation of condensation under the plastic covering. If this is the case, use a covering that is able to "breathe," such as a canvas tarpaulin.

ENGINE TUNE-UP PROCEDURE

The following list of procedures is rather extensive for a simple tune-up. Normally one would just check the condition of the spark plug, points, condenser, and wiring, make the necessary adjustments to these components and the carburetor, maybe change the oil, and service the carburetor if needed.

However if the following is performed, you will either be sure that the engine is functioning properly or you will know what major repairs should be made. In other words the engine is going to run well or you will find the cause of any problems.

1. Remove the air cleaner and check for the proper servicing.

2. Check the oil level and drain the crankcase. Clean the fuel tank and lines if it is separate from the carburetor.

3. Remove the blower housing and inspect the rope, rewind assembly, and starter clutch of the starter mechanism.

4. Clean the cooling fins and the entire engine. Turn the flywheel to check compression.

5. Remove the carburetor and disassemble and inspect it for wear or damage. Wash it in solvent, replace parts as necessary, and assemble. Set the initial adjustments.

6. Inspect the crossover tube or the intake elbow for damaged gaskets.

7. Check the governor blade, linkage, and spring for damage or wear; if it is mechanical, check the linkage adjustment.

8. Remove the flywheel and check for seal leakage, both on the flywheel and power take off sides. Check the flywheel key for wear and damage.

9. Remove the breaker cover and check for proper sealing.

10. Inspect the breaker points and condenser. Replace or clean and adjust them. Check the plunger or the cam.

11. Check the coil and inspect all wires for breaks or damaged insulation. Be sure the lead wires do not touch the flywheel. Check the stop switch and the lead.

12. Replace the breaker cover, using sealer where the wires enter.

13. Install the flywheel and time the ignition if necessary. Set the air gap and check for ignition spark.

14. Remove the cylinder head, check the gasket, remove the spark plug, clean off the carbon, and inspect the valves for proper seating.

15. Replace the cylinder head, using a new gasket, torque it to the proper specification, and set the spark plug gap or replace the plug if necessary.

16. Replace the oil and fuel and check the muffler for restrictions or damage.

17. Adjust the remote control linkage and cable, if used, for correct operation.

18. Service the air cleaner and check the gaskets and element for damage.

19. Run the engine and adjust the idle mixture and high speed mixture of the carburetor.

GENERAL MAINTENANCE OF A SMALL ENGINE

Before starting the engine, fill the crankcase and the air cleaner with the proper oil and fill the gasoline tank. Never try to fill the fuel tank of an engine that is running,

and if the engine is still hot from running, allow it to cool down before refueling it.

Use a good grade, clean, fresh, lead free or leaded regular grade automotive gasoline. The use of highly leaded gasoline (high octane) should be avoided, as it causes deposits on the valves and valve seats, spark plugs, and the cylinder head, thus shortening engine life.

Any high quality detergent oil having the American Petroleum Institute classification "For Service SC or SD or MS" can be used. Detergent oils keep the engine cleaner by retarding the formation of gum and varnish deposits. Do not use any oil additives. In the summer (above 40° F.) use SAE 30 weight oil. If that is not available, use SAE 10W-30 or SAE 10W-40 weight oil. In the winter (under 40° F.) use SAE 5W-20 or SAE 5W-30 weight oil. If neither of these is available, use SAE 10W or SAE 10W-30 weight oil. If the engine is operated in ambient temperatures that are below 0° F., use SAE 10W or SAE 10W-30 weight oil diluted 10% with kerosene.

The oil should be changed after each 25 hours of service or engine operation, and more often under dirty or dusty operating conditions. In normal running of any engine, small particles of metal from the cylinder walls, pistons and bearings will gradually work into the oil. Dust particles from the air also get into the oil. If the oil is not changed regularly, these foreign particles cause increased friction and a grinding action which shorten the life of the engine. Fresh oil also assists in cooling the engine, for old oil gradually becomes thick and cannot dissipate the heat fast enough. Old oil will also gradually lose its lubricating properties.

The air cleaner should be serviced every 25 hours of engine operation.

Keep the engine clean and free from all grass particles, chaff and oil soaked dirt. This is so that the engine's cooling system will be able to operate at full efficiency. Continued operation with a clogged cooling system may cause severe overheating and possible engine damage. Remove the blower housing often to be sure the passages are free of any debris. Don't forget to replace the housing on the engine. This should be included in your regular maintenance schedule.

2 · Briggs and Stratton

Engine Identification

The Briggs and Stratton model designation system consists of up to a six digit number. It is possible to determine most of the important mechanical features of the engine by merely knowing the model number. An explanation of what each number means is given below.

1. The first one or two digits indicate the cubic inch displacement (cid).

2. The first digit after the displacement indicates the basic design series, relating to cylinder construction, ignition and general configuration.

3. The second digit after the displacement indicates the position of the crankshaft and the type of carburetor the engine has.

4. The third digit after the displacement indicates the type of bearings and whether or not the engine is equipped with a reduction gear or auxiliary drive.

5. The last digit indicates the type of starter.

CUBIC INCH DISPLACEMENT	FIRST DIGIT AFTER DISPLACEMENT BASIC DESIGN SERIES	SECOND DIGIT AFTER DISPLACEMENT CRANKSHAFT, CARBURETOR GOVERNOR	THIRD DIGIT AFTER DISPLACEMENT BEARINGS, REDUCTION GEARS & AUXILIARY DRIVES	FOURTH DIGIT AFTER DISPLACEMENT TYPE OF STARTER
6	0	0 -	0 - Plain Bearing	0 - Without Starter
8	1	1 - Horizontal Vacu-Jet	1 - Flange Mounting Plain Bearing	1 - Rope Starter
9	2	2 - Horizontal Pulsa-Jet	2 - Ball Bearing	2 - Rewind Starter
10	3	3 - Horizontal Flo-Jet (Pneumatic Governor)	3 - Flange Mounting Ball Bearing	3 - Electric - 110 Volt, Gear Drive
13	4	4 - Horizontal Flo-Jet (Mechanical Governor)	4 -	4 - Elec. Starter-Generator - 12 Volt, Belt Drive
14	5			
17	6	5 - Vertical Vacu-Jet	5 - Gear Reduction (6 to 1)	5 - Electric Starter Only - 12 Volt, Gear Drive
19	7			
20	8	6 -	6 - Gear Reduction (6 to 1) Reverse Rotation	6 - Wind-up Starter
23	9			
24		7 - Vertical Flo-Jet	7 -	7 - Electric Starter, 12 Volt Gear Drive, with Alternator
30		8 -	8 - Auxiliary Drive Perpendicular to Crankshaft	8 - Vertical-pull Starter
32		9 - Vertical Pulsa-Jet	9 - Auxiliary Drive Parallel to Crankshaft	

EXAMPLES

To identify Model 100202:

10	0	2	0	2
10 Cubic Inch	Design Series 0	Horizontal Shaft - Pulsa-Jet Carburetor	Plain Bearing	Rewind Starter

Similarly, a Model 92996 is described as follows:

9	2	9	9	6
9 Cubic Inch	Design Series 2	Vertical Shaft - Pulsa-Jet Carburetor	Auxiliary Drive Parallel to Crankshaft	Wind-up Starter

Briggs and Stratton model numbering system

Air Cleaners

A properly serviced air cleaner protects the engine from dust particles that are in the air. When servicing an air cleaner, check the air cleaner mounting and gaskets for worn or damaged mating surfaces. Replace any worn or damaged parts to prevent dirt and dust from entering the engine through openings caused by improper sealing. Straighten or replace any bent mounting studs.

Servicing

OIL FOAM AIR CLEANERS

Clean and re-oil the air cleaner element every 25 hours of operation under normal operating conditions. The capacity of the oil-foam air cleaner is adequate for a full season's use without cleaning. Under very dusty conditions, clean the air cleaner every few hours of operation.

The oil-foam air cleaner is serviced in the following manner:

1. Remove the screw that holds the halves of the air cleaner shell together and retains it to the carburetor.

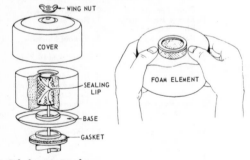

Oil foam air cleaner

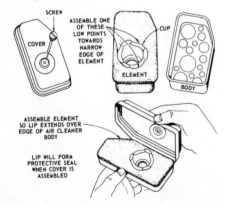

Oil foam air cleaner

2. Remove the air cleaner carefully to prevent dirt from entering the carburetor.

3. Take the air cleaner apart (split the two halves).

4. Wash the foam in kerosene or liquid detergent and water to remove the dirt.

5. Wrap the foam in a clean cloth and squeeze it dry.

6. Saturate the foam in clean engine oil and squeeze it to remove the excess oil.

7. Assemble the air cleaner and fasten it to the carburetor with the attaching screw.

OIL BATH AIR CLEANER

Pour the old oil out of the bowl. Wash the element thoroughly in solvent and squeeze it dry. Clean the bowl and refill it with the same type of oil used in the crankcase.

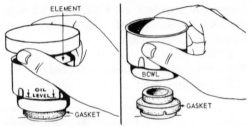

Oil bath air cleaner

DRY ELEMENT AIR CLEANER

Remove the element of the air cleaner and tap (top and bottom) it on a flat surface or wash it in non-sudsing detergent and flush it from the inside until the water coming out is clear. After washing, air dry the element thoroughly before reinstalling it on the engine. NEVER OIL A DRY ELEMENT.

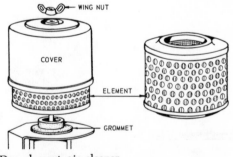

Dry element air cleaner

HEAVY DUTY AIR CLEANER

Clean and re-oil the foam pre-cleaner at three month intervals or every 25 hours, whichever comes first.

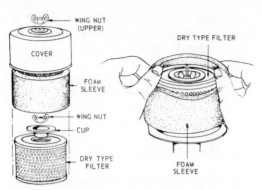

WING NUT
(UPPER)

COVER

FOAM
SLEEVE

WING NUT

CUP

DRY TYPE
FILTER

DRY TYPE FILTER

FOAM
SLEEVE

Briggs and Stratton heavy duty air cleaner

Clean the paper element every year or 100 hours, whichever comes first. Use the dry element procedure for cleaning the paper element of the heavy duty air cleaner.

Use the oil foam cleaning procedure to clean the foam sleeve of the heavy duty air cleaner.

If the engine is operated under very dusty conditions, clean the air cleaner more often.

Carburetors

There are three types of carburetors used on Briggs and Stratton engines. They are the Pulsa-Jet, Vacu-Jet and Flo-Jet. The first two types have three models each and the Flo-Jet has two versions.

Before removing any carburetor for repair, look for signs of air leakage or mounting gaskets that are loose, have deteriorated, or are otherwise damaged.

Note the position of the governor springs, governor link, remote control, or other attachments to facilitate reassembly. Be careful not to bend the links or stretch the springs.

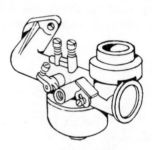

The two types of Flo-Jet carburetors

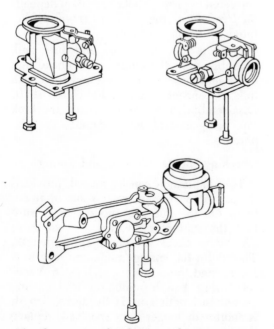

The three types of Pulsa-Jet carburetors

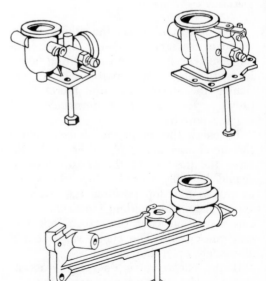

The three types of Vacu-Jet carburetors

AUTOMATIC CHOKE

All 92000 model engines built since August 1968 have an automatic choke system.

The automatic choke operates in conjunction with engine vacuum, similar to the Pulsa-Jet fuel pump.

A diaphragm under the carburetor is connected to the choke shaft by a link. A calibrated spring under the diaphragm holds the choke closed when the engine is not running. Upon starting, vacuum created during the intake stroke is routed to the bottom of the diaphragm through a calibrated passage, thereby opening the choke.

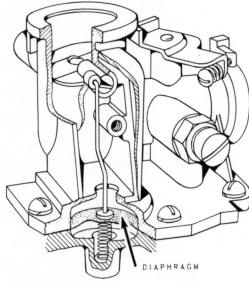

Automatic choke system

This system also has the ability to respond in the same manner as an accelerator pump. As speed decrease during heavy loads, the choke valve partially closes, enriching the air/fuel mixture, thereby improving low speed performance and lugging power.

To check the automatic choke, remove the air cleaner and replace the stud. Observe the position of the choke valve; it should be fully closed. Move the speed control to the stop position; the governor spring should be holding the throttle in a closed position. Give the starter rope several quick pulls. The choke valve should alternately open and close.

If the choke valve does not react as stated in the previous paragraph, the carburetor will have to be disassembled to determine the problem. Before doing so, however, check the following items so you know what to look for:

Engine is under-choked.

1. Carburetor is adjusted too lean.
2. The fuel pipe check valve is inoperative (Vacu-Jet only).
3. The air cleaner stud is bent.
4. The choke shaft is sticking due to dirt.
5. The choke spring is too short or damaged.
6. The diaphragm is not preloaded.

Engine is over-choked.

1. Carburetor is adjusted too rich.
2. The air cleaner stud is bent.
3. The choke shaft is sticking due to dirt.
4. The diaphragm is ruptured.
5. The vacuum passage is restricted.
6. The choke spring is distorted or stretched.
7. There is gasoline or oil in the vacuum chamber.
8. There is a leak between the link and the diaphragm.
9. The diaphragm was folded during assembly, causing a vacuum leak.
10. The machined surface on the tank top is not flat.

Repairing the Automatic Choke

Inspect the automatic choke for free operation. Any sticking problems should be corrected as proper choke operation depends on freedom of the choke to travel as dictated by engine vacuum.

Remove the carburetor and fuel tank assembly from the engine. The choke link cover may now be removed and the choke link disconnected from the choke shaft. Disassemble the carburetor from the tank top, being careful not to damage the diaphragm.

Checking the Diaphragm and Spring

The diaphragm can be reused, provided it has not developed wear spots or punctures. On the Pulsa-Jet models, make sure that the fuel pump valves are not damaged. Also check the choke spring length. The Pulsa-Jet spring minimum length is $1\frac{1}{8}$ in. and the maximum is $1\frac{7}{32}$ in. Vacu-Jet spring length minimum is $\frac{15}{16}$ in., maximum length 1 in. If the spring length is shorter or longer than specified, replace the diaphragm and the spring.

Checking the Tank Top

The machined surface on the top of the tank must be flat in order for the diaphragm to provide an adequate seal between the carburetor and the tank. If the machined surface on the tank is not flat, it is possible for gasoline to enter the vacuum chamber by passing between the machined surface and the diaphragm. Once fuel has entered the vacuum chamber, it can move through the vacuum passage and into the carburetor. The flatness of the machined surface on the tank top can be checked by using a straightedge and a feeler gauge. The surface should not vary more than 0.002 in. Replace the tank if a 0.002 in. feeler gauge can be passed under the straightedge.

If a new diaphragm is installed, assemble the spring to the replacement dia-

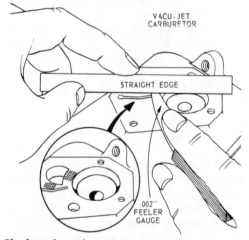

Checking the tank top for warpage

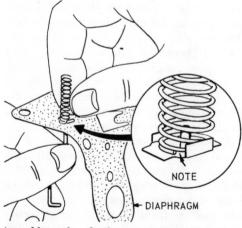

Assembling the diaphragm spring to the new diaphragm

phragm, taking care not to bend or distort the spring.

Place the diaphragm on the tank surface, positioning the spring in the spring pocket.

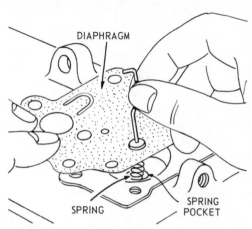

Installing the diaphragm and spring into the spring pocket

Place the carburetor on the diaphragm ensuring that the choke link and diaphragm are properly aligned between the carburetor and the tank top. On Pulsa-Jet models, place the pump spring and cap on the diaphragm over the recess or pump chamber in the fuel tank. Thread in the carburetor mounting screws to about two threads. Do not tighten them. Close the choke valve and insert the choke link into the choke shaft.

Remove the air cleaner gasket, if it is in place, before continuing. Insert a ⅜ in. bolt or rod into the carburetor air horn. With the bolt in position, tighten the carburetor mounting screws in a staggered sequence. Please note that the insertion of

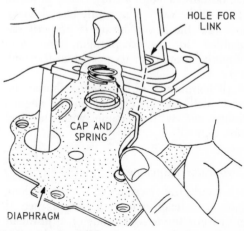

Positioning the diaphragm on top of the fuel tank

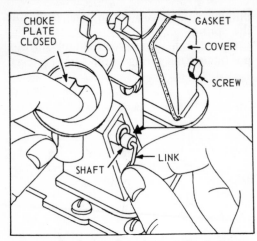

Inserting the choke link into the choke shaft

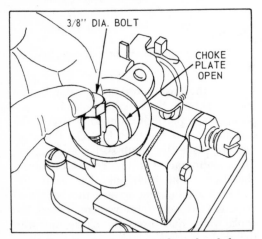

Pre-loading the diaphragm to adjust the choke

the ⅜ in. bolt opens the choke to an over-center position, which preloads the diaphragm.

Remove the ⅜ in. bolt. The choke valve should now move to a fully closed position. If the choke valve is not fully closed, make sure that the choke spring is properly assembled to the diaphragm, and also properly inserted in its pocket in the tank top.

All carburetor adjustments should be made with the air cleaner on the engine. Adjustment is best made with the fuel tank half full.

Carburetor Adjustment

1. Start the engine and run it long enough to reach operating temperature. If the carburetor is so far out of adjustment that it will not start, close the needle valve by turning it clockwise. Then open the needle valve 1½ turns counterclockwise.

2. Move the control so that the engine runs at normal operating speed. Turn the needle valve clockwise until the engine starts to lose speed because of too lean a mixture. Then slowly turn the needle valve counterclockwise and out past the point of smoothest operation until the engine just begins to run unevenly because of too rich a mixture. Turn the needle back clockwise to the midpoint between the rich and lean mixture extremes. This should be where the engine operates smoothest. The final adjustment of the needle valve should be slightly on the rich side (counterclockwise) of the mid-point.

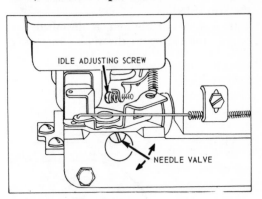

Carburetor adjustment screws

3. Move the engine control to the slow position and turn the idle adjusting screw until a fast idle of about 1750 rpm is obtained. If the engine idles at a speed lower than 1750 rpm, it may not accelerate properly. It is not practical to attempt to obtain acceleration from speeds below 1750 rpm, because the mixture which would be required would be too rich for normal operating speeds.

4. To check the idle adjustment, move the engine control from slow to fast speed. The engine should accelerate smoothly. If the engine tends to stall or die out, increase the idle speed or readjust the carburetor, usually to a slightly richer mixture.

Flooding can occur if the engine is tipped at an angle for a prolonged period of time, if the engine is cranked repeatedly with the spark plug wire disconnected, or if the carburetor mixture is too rich.

In case of flooding, move the governor control to the stop position and pull the starter rope at least six times.

When the control is placed in the stop position, the governor spring holds the throttle in a closed idle position. Cranking the engine with a closed throttle creates a higher vacuum which opens the choke rapidly, permitting the engine to clear itself of excess fuel.

Then move the control to the fast position and start the engine. If the engine continues to flood, lean the carburetor needle valve by about $\frac{1}{8}$–$\frac{1}{4}$ of a turn clockwise.

PULSA-JET AND VACU-JET (MODEL SERIES 82000, 92000 ONLY)

Models 82500 and 92500 have a Vacu-Jet carburetor and Models 82900 and 92900 have a Pulsa-Jet carburetor.

Rebuilding

1. Remove the carburetor and fuel tank assembly from the engine by removing the two attaching bolts.

2. Disconnect the governor link at the throttle, leaving the governor link and the governor spring hooked to the governor blade and control lever.

3. Slip the carburetor and tank assembly off of the engine.

4. Remove the carburetor from the tank. Always remove all nylon and rubber parts if the carburetor is soaked in solvent.

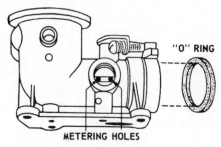

Remove the O-ring and inspect the metering valve

5. Remove the O-ring and discard it. Remove and inspect the needle valve, packing, and seat.

6. Metering holes in the carburetor body should be cleaned with solvent and compressed air. Do not clean the holes with a pin or a length of wire because of the danger of altering their size.

7. Remove the choke parts on models 82500 and 82900 by pulling the nylon choke shaft sideways to separate the choke shaft from the choke valve. On the 92500

and 92900, remove the choke parts by first disconnecting the choke return spring at the pin in the carburetor body. Then pull the nylon choke shaft sideways to separate the choke shaft from the choke valve.

8. If the choke valve is heat-sealed to the choke shaft, loosen it by sliding a sharp pointed tool along the edge of the choke shaft. Do not re-seal parts on assembly.

9. When replacing the choke valve and shaft, install the choke valve so the poppet valve spring is visible when the valve is in full choke position.

MODEL 82000 AND 92000 VARIANCES

The nylon fuel pipe is threaded into the carburetor body. Use a socket to remove and replace it. Be careful not to over-tighten it and do not use any sealer.

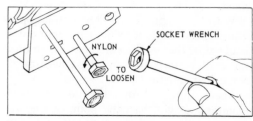

Removing the nylon fuel pipes

The Pulsa-Jet diaphragm also serves as a gasket between the carburetor and the tank. Inspect the diaphragm for punctures, wrinkles, and wear. Replace it if it is damaged in any way.

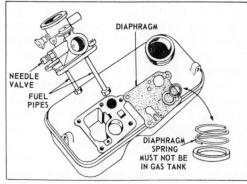

Removal and inspection of a Pulsa-Jet diaphragm

To assemble the carburetor to the tank, first position the diaphragm on the tank. Then place the spring cap and spring on the diaphragm. Install the carburetor, tightening the mounting screws evenly to avoid distortion.

To install the carburetor and tank assembly onto the engine, make sure that the governor link is hooked to the governor blade. Connect the link to the throttle and slip the carburetor into place. Align the carburetor with the intake tube and breather tube grommet. Hold the choke lever in the open position so it does not catch on the control plate. Be sure the O-ring in the carburetor does not distort when fitting the carburetor to the intake tube. Install the mounting bolts.

Adjustment

Adjust the carburetor with the air cleaner installed and the fuel tank half full.

Turn the needle valve clockwise to close it. Then open it about 1½ turns. This will permit the engine to be started and warmed up before making the final adjustment.

With the engine running at normal operating speed (about 3000 rpm without a load) turn the needle valve clockwise until the engine starts to lose speed because of a too lean mixture.

Then slowly turn the needle valve counterclockwise past the point of smoothest operation, until the engine just begins to run unevenly. This mixture will give the best performance under a load.

Hold the throttle in the idle position. Turn the idle speed adjusting screw until a fast idle is obtained (about 1750 rpm).

Test the engine under full load. If the engine tends to stall or die out, it usually indicates that the mixture is slightly lean and it may be necessary to open the needle valve slightly to provide a richer mixture. This slightly richer mixture may cause a slight unevenness in idling.

The breather tube and fuel intake tube thread into the cylinder on the model 82500 and 82900 engines. The fuel intake tube is bolted to the cylinder on the model 92500 and 92900 engines. Check for a good fit to prevent any air leaks or dirt entry. The fuel intake tube must not be distorted at the point where the carburetor O-ring fits or air leaks will occur.

PULSA-JET CARBURETOR

Throttle Plate Removal

Cast throttle plates are removed by backing off the idle speed adjustment screw until the throttle clears the retaining lug on the carburetor housing.

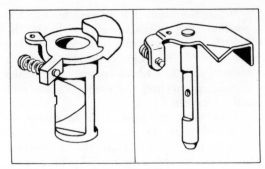

The two types of throttle shafts

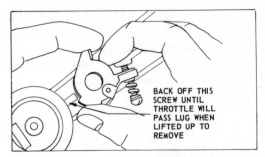

Removing the cast throttle shafts

Stamped throttles are removed by using a phillips screwdriver to remove the throttle valve screw. After removal of the valve, the throttle may be lifted out. Installation is the reverse of removal.

Some carburetors may have a spiral in the carburetor bore. To remove it, fasten

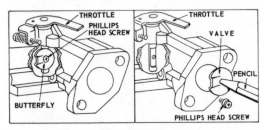

Removing the throttle plate

the carburetor in a vise about ½ in. below the top of the jaws. Grasp the spiral firmly with a pair of pliers. Place a screwdriver under the edge of the pliers. Using the edge of the vise, push down on the screwdriver to pry out the spiral. When installing the spiral, keep the top flush, or $\frac{1}{32}$ in. below the carburetor flange, and parallel with the fuel tank mounting face.

Fuel Pipe

Check balls are not used in these fuel pipes. The screen housing or pipe must be replaced if the screen cannot be satisfactorily cleaned. The long pipe supplies fuel

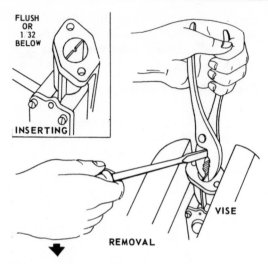

Removing and installing the spiral

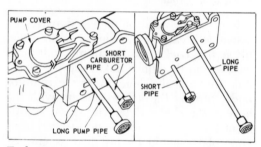

Fuel pipes

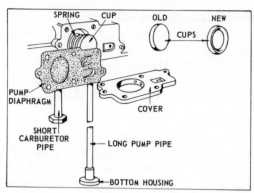

Removal and inspection of the pump cover diaphragm from a Pulsa-Jet carburetor

from the tank to the pump. The short pipe supplies fuel from the tank cup to the carburetor. Fuel pipes are nylon or brass. Nylon pipes are removed and installed by using a socket, or open-end wrench.

NOTE: *Where brass pipes are used, replace only the screen housing. The housing is driven off the pipe with a screwdriver with the pipe held in a vise. The new housing is installed by lightly tapping it onto the pipe with a soft hammer.*

Needle Valve and Seat

Remove the needle valve to inspect it. If the carburetor is gummy or dirty, remove the seat to allow better cleaning of the metering holes. Do not insert pins or wires in the metering holes. Use solvent or compressed air.

Pump

Remove the fuel pump cover, diaphragm, spring, and cup. Inspect the diaphragm for punctures, cracks, and fatigue. Replace it if damaged. On early models, the spring cap is solid; on later models, the cap has

a hole in it. The new style supersedes the old style. When installing the pump cover, tighten the screws evenly to insure a good seal.

Choke-A-Matic (Except 100900 Models)

To remove the choke link, remove the speed adjustment lever and stop switch insulator plate. Remove the speed adjustment lever from the choke link, then pull out the choke link through the hole in the choke slide.

Replace worn or damaged parts. To assemble, slip the washers and spring over the choke link. Hook the choke link through the hole in the choke slide. Place the other end of the choke link through the hole in the speed adjustment lever and mount the lever and stop switch insulator plate to the carburetor.

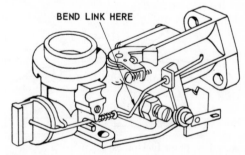

Adjustment of the Choke-a-Matic choke linkage

To check the operation of the Choke-A-Matic linkage, move the speed adjustment lever to the choke position. If the choke slide does not fully close, use flat nose pliers to bend the choke link, The speed adjustment lever must make good contact against the stop switch when it is moved to the stop position.

VACU-JET CARBURETORS

Vacu-Jet carburetors are removed from the engine together with the fuel tank as one unit. The throttle plates are removed and installed in the same manner as the throttles in the Pulsa-Jet carburetors.

Fuel Pipe

The fuel pipe contains a check ball and a fine mesh screen. To function properly, the screen must be clean and the check ball free. Replace the pipe if the screen and ball cannot be satisfactorily cleaned in carburetor cleaner.

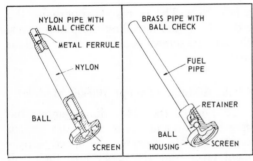

The two types of fuel pipes

NOTE: *Do not leave the carburetor in the cleaner for more than ½ hour without removing all nylon parts. Nylon fuel pipes are removed and replaced with a 9⁄16 in. socket. Brass fuel pipes are removed by clamping the pipe in a vise and prying out the pipe with two screwdrivers.*

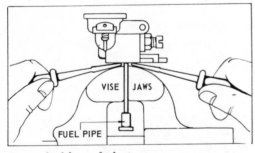

Removal of brass fuel pipes

To install the brass fuel pipes, remove the throttle, if necessary, and place the carburetor and pipe in a vise. Press the pipe into the carburetor until it projects $2\frac{9}{32}-2\frac{5}{16}$ in. from the carburetor face.

Needle Valve and Seat

Remove the needle valve assembly to inspect it. If the carburetor is gummy or dirty, remove the seat to allow better cleaning of the metering holes. Do not clean the metering holes with a pin or a length of wire.

Choke-A-Matic Linkage

To remove the choke link, remove the speed adjustment lever and the stop switch insulator plate. Work the link out through the hole in the choke slide.

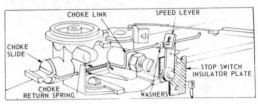

Adjustment of a Choke-a-Matic choke linkage on a Vacu-Jet carburetor

Replace all worn or damaged parts. To assemble a carburetor using a choke slide, place the choke return spring and three washers on the choke link. Push the choke link through the hole in the carburetor body, turning the link to line up with the hole in the choke slide. The speed adjustment lever screw and the stop switch insulator plate should be installed as one assembly after placing the choke link through the end of the speed adjustment lever.

Adjustment of the Choke-A-Matic Linkage

To check the operation of the choke linkage, move the speed adjustment lever to the choke position. If the choke slide does not fully close, bend the choke link. The speed adjustment lever must make good contact against the stop switch.

Install the carburetor and adjust it in the same manner as the Pulsa-Jet carburetor.

TWO PIECE FLO-JET CARBURETORS (LARGE AND SMALL LINE)

Checking the Upper Body for Warpage

With the carburetor assembled and the body gasket in place, try to insert a 0.002 in. feeler gauge between the upper and lower bodies at the air vent boss, just below the idle valve. If the gauge can be inserted, the upper body is warped and should be replaced.

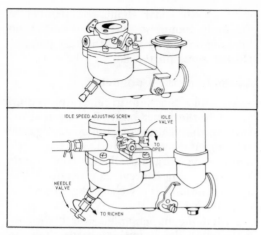

Two piece Flo-Jet carburetor

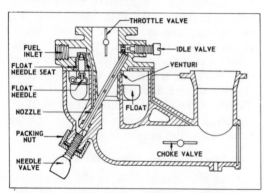

Cutaway view of a two piece Flo-Jet carburetor

Checking the Throttle Shaft and Bushings

Wear between the throttle shaft and bushings should not exceed 0.010 in. Check the wear by placing a short iron bar on the upper carburetor body so that it just fits under the throttle shaft. Measure the distance with a feeler gauge while holding the shaft down and then holding it up. If the difference is over 0.010 in., either the upper body should be rebushed, the throttle shaft replaced, or both. Wear on the

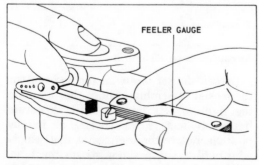

Checking throttle shaft wear with a feeler gauge

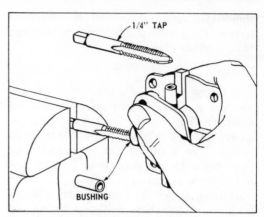

Removing the throttle shaft bushing

throttle shaft can be checked by comparing the worn and unworn portions of the shaft. To replace the bushings, remove the throttle shaft using a thin punch to drive out the pin which holds the throttle stop to the shaft; remove the throttle valve, then pull out the shaft. Place a ¼ in. x 20 tap or an E-Z Out ® in a vise. Turn the carburetor body so as to thread the tap or E-Z Out ® into the bushings enough to pull the bushings out of the body. Press the new bushings into the carburetor body with a vise. Insert the throttle shaft to be sure it is free in the bushings. If not, run a size $7/32$ in. drill through both bushings to act as a line reamer. Install the throttle shaft, valve, and stop.

Disassembly of the Carburetor

1. Remove the idle valve.
2. Loosen the needle valve packing nut.
3. Remove the packing nut and needle valve together. To remove the nozzle, use a narrow, blunt screwdriver so as not to damage the threads in the lower carburetor body. The nozzle projects diagonally into a recess in the upper body and must be removed before the upper body is separated from the lower body, or it may be damaged.
4. Remove the screws which hold the upper and lower bodies together. A pin holds the float in place.
5. Remove the pin to take out the float valve needle. Check the float for leakage. If it contains gasoline or is crushed, it must be replaced. Use a wide, proper fitting screwdriver to remove the float inlet seat.
6. Lift the venturi out of the lower body. Some carburetors have a welch plug. This

should be removed only if necessary to remove the choke plate. Some carburetors have a nylon choke shaft.

Repair

Use new parts where necessary. Always use new gaskets. Carburetor repair kits are available. Tighten the inlet seat with the gasket securely in place, if used. Some float valves have a spring clip to connect the float valve to the float tang. Others are nylon with a stirrup which fits over the float tang. Older float valves and engines with fuel pumps have neither a spring nor a stirrup.

A viton tip float valve is used in later models of the large, two-piece Flo-Jet carburetor. The seat is pressed into the upper body and does not need replacement unless it is damaged.

Replacing the Pressed-In Float Valve Seat

Clamp the head of a #93029 self threading screw in a vise. Turn the carburetor body to thread the screw into the seat. Continue turning the carburetor body, drawing out the seat. Leave the seat fastened to the screw. Insert the new seat #230996 into the carburetor body. The seat has a starting lead.

NOTE: *If the engine is equipped with a fuel pump, install a #231019 seat. Press the new seat flush with the body using the screw and old seat as a driver. Make sure that the seat is not pressed below the body surface or improper float-to-float valve contact will occur. Install the float valve.*

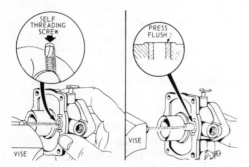

Replacing the float valve seat

Checking the Float Level

With the body gasket in place on the upper body and the float valve and float

installed, the float should be parallel to the body mounting surface. If not, bend the tang on the float until they are parallel. Do not press on the flat to adjust it.

Assembly of the Carburetor

Assemble the venturi and the venturi gasket to the lower body. Be sure that the holes in the venturi and the venturi gasket are aligned. Some models do not have a removable venturi. Install the choke parts and welch plug if previously removed. Use a sealer around the welch plug to prevent entry of dirt.

Fasten the upper and lower bodies together with the mounting screws. Screw in the nozzle with a narrow, blunt screwdriver, making sure that the nozzle tip enters the recess in the upper body. Tighten the nozzle securely. Screw in the needle valve and idle valve until they just seat. Back off the needle valve 1½ turns. Do not tighten the packing nut. Back off the idle valve ¾ of a turn. These settings are about correct. Final adjustment will be made when the engine is running.

Two-Piece Flo-Jet Automatic Choke

Hold the choke shaft so the thermostat lever is free. At room temperature (68° F), the screw in the thermostat collar should be in the center of the stops. If not, loosen the stop screw and adjust the screw.

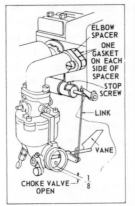

Automatic choke adjustment

Loosen the set screw on the lever of the thermostat assembly. Slide the lever to the right or left on the shaft to ensure free movement of the choke link in any position. Rotate the thermostat shaft clockwise until the stop screw strikes the tube. Hold it in position and set the lever on the thermostat shaft so that the choke valve will be held open about ⅛ in. from a closed position. Then tighten the set screw in the lever.

Rotate the thermostat shaft counter-clockwise until the stop screw strikes the opposite side of the tube. Then open the choke valve manually until it stops against the top of the choke link opening. The choke valve should now be open approximately ⅛ in. as before.

Check the position of the counterweight lever. With the choke valve in a wide open position (horizontal) the counterweight lever should also be in a horizontal position with the free end toward the right.

Operate the choke manually to be sure that all parts are free to move without binding or rubbing in any position.

ONE-PIECE FLO-JET CARBURETOR

The large, one-piece Flo-Jet carburetor has its high speed needle valve below the float bowl. All other repair procedures are similar to the small, one-piece Flo-Jet carburetor.

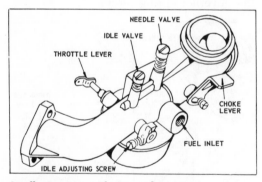

Small one piece Flo-Jet carburetor

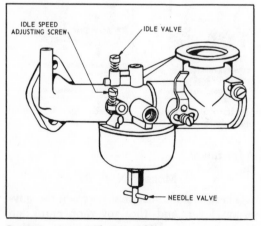

Large one piece Flo-Jet carburetor

Disassembly

1. Remove the idle and needle valves.
2. Remove the carburetor bowl screw. A pin holds the float in place.
3. Remove the pin to take off the float and float valve needle. Check the float for leakage. If it contains gasoline or is crushed, it must be replaced. Use a screwdriver to remove the carburetor nozzle. Use a wide, heavy screwdriver to remove the float valve seat, if used.

If it is necessary to remove the choke valve, venturi throttle shaft, or shaft bushings, proceed as follows.

1. Pry out the welch plug.
2. Remove the choke valve, then the shaft. The venturi will then be free to fall out after the choke valve and shaft have been removed.
3. Check the shaft for wear. (Refer to the "Two-Piece Flo-Jet Carburetor" section for checking wear and replacing bushings.)

Repair of the Carburetor

Use new parts where necessary. Always use new gaskets. Carburetor repair kits are available. If the venturi has been removed, install the venturi first, then the carburetor nozzle jets. The nozzle jet holds the venturi in place. Replace the choke shaft and valve. Install a new welch plug in the carburetor body. Use a sealer to prevent dirt from entering.

A viton tip float valve is used in the large, one-piece Flo-Jet carburetor. The seat is pressed in the upper carburetor body and does not need replacement unless it is damaged. Replace the seat in the same manner as for the two-piece Flo-Jet carburetor.

Checking the Float Level

With the body gasket in place on the upper body and float valve and the float installed, the float should be parallel to the body mounting surface. If not, bend the tang on the float until they are parallel. Do not press on the float.

Install the float bowl, idle valve, and needle valve. Turn in the needle valve and the idle valve until they just seat. Open the needle valve 2½ turns and the idle valve 1½ turns. On the large carburetors with the needle valve below the float bowl,

Float Level Chart

Carburetor Number	Float Setting (in.)
2712-S	$1\frac{9}{64}$
2713-S	$1\frac{9}{64}$
2714-S	$\frac{1}{4}$
*2398-S	$\frac{1}{4}$
2336-S	$\frac{1}{4}$
2336-SA	$\frac{1}{4}$
2337-S	$\frac{1}{4}$
2337-SA	$\frac{1}{4}$
2230-S	$1\frac{7}{64}$
2217-S	$1\frac{1}{64}$

* When resilient seat is used, set float level at $\frac{9}{32} \pm \frac{1}{64}$.

open the needle valve and the idle valve $1\frac{1}{8}$ turns.

These settings will allow the engine to start. Final adjustment should be made when the engine is running and has warmed up to operating temperature. See the "Two-Piece Flo-Jet Carburetor" adjustment procedure.

Governors

The purpose of a governor is to maintain, within certain limits, a desired engine speed even though the load may vary.

AIR VANE GOVERNORS

The governor spring tends to open the throttle. Air pressure against the air vane tends to close the throttle. The engine speed at which these two forces balance is called the governed speed. The governed speed can be varied by changing the governor spring tension.

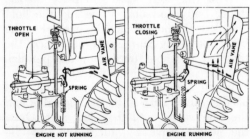

Air vane governor installed on horizontal crankshaft engine

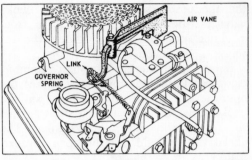

Air vane governor installed on vertical crankshaft engine

Worn linkage or damaged governor springs should be replaced to insure proper governor operation. No adjustment is necessary.

MECHANICAL GOVERNORS

The governor spring tends to pull the throttle open. The force of the counterweights, which are operated by centrifugal force, tends to close the throttle. The engine speed at which these two forces balance is called the governed speed. The governed speed can be varied by changing the governor spring tension.

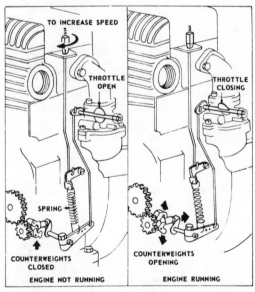

Mechanical governor installed on horizontal crankshaft engine

Adjustment

MODELS N, 6, 8

There is no adjustment between the governor lever and the governor crank on these models. However, governor action can be changed by inserting the governor link or spring in different holes of the governor and throttle levers. In general, the closer to the pivot end of the lever, the smaller the difference between load and no-load engine speed. The engine will begin to "hunt" if the spring is brought too close to the pivot point. The farther the spring is from the pivot end, the tendency to hunt will decrease, but the speed drop will be greater as the load increases. If the governor speed is lowered, the spring can usually be moved closer to the pivot. The standard setting is the 4th hole from the pivot point.

Models 6B, 8B, 60000, 80000, 140000

Loosen the screw which holds the governor lever to the governor shaft. Turn the governor lever counterclockwise until the carburetor throttle is wide open. With a screwdriver, turn the governor shaft counterclockwise as far as it will go. Tighten the screw which holds the governor lever to the governor shaft.

Cast Iron Models
9, 14, 19, 190000, 200000

Loosen the screw which holds the governor lever to the governor shaft. Push the lever counterclockwise as far as it will go. Hold it in this position and turn the governor shaft counterclockwise as far as it will go. This can be done with a screwdriver. Securely tighten the screw that holds the governor lever to the shaft.

Aluminum Models
100000, 130000, 140000, 170000, 190000

Vertical and horizontal shaft engine governors are adjusted by setting the control lever in the high speed position. Loosen the nut on the governor lever. Turn the governor shaft clockwise to the end of its travel. Tighten the nut. The throttle must be wide open. Check to see if the throttle can be moved from idle to wide open without binding.

Adjusting Top No Load Speed

Set the control lever to the maximum speed position with the engine running. Bend the spring anchor tang to get the desired top speed.

Ignition Systems

All Briggs and Stratton engines have magneto ignition systems. Three types are used: Flywheel Type—Internal Breaker Flywheel Type—External Breaker, and Magna-Matic.

FLYWHEEL TYPE—INTERNAL BREAKER

This ignition system has the magneto located on the flywheel and the breaker points located under the flywheel.

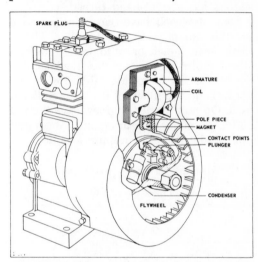

Flywheel magneto ignition with internal breaker points and external armature

The flywheel is located on the crankshaft with a soft metal key. It is held in place by a nut or starter clutch. The flywheel key must be in good condition to insure proper location of the flywheel for ignition timing. Do not use a steel key under any circumstances. Use only a soft metal key, as originally supplied.

The keyway in both flywheel and crankshaft should not be distorted. Flywheels are made of aluminum, zinc, or cast iron.

Flywheel, Nut, and/or Starter Clutch

Removal and Installation

Place a block of wood under the flywheel fins to prevent the flywheel from turning while you are loosening the nut or starter clutch. Be careful not to bend the flywheel. There are special flywheel holders available for this purpose.

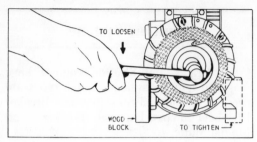

Removing the flywheel

On rope starter engines, the ½ in. flywheel nut has a left-hand thread and the ⅝ in. nut has a right-hand thread. The starter clutch used on rewind and wind-up starters has a right-hand thread.

Some flywheels have two holes provided for the use of a flywheel puller. Use a small gear puller or automotive steering wheel puller to remove the flywheel if a flywheel puller is not available. Be careful not to bend the flywheel if a gear puller is used. On rope starter engines leave the nut on for the puller to bear against. Small cast iron flywheels do not require a puller.

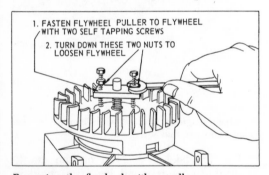

Removing the flywheel with a puller

Install the flywheel in the reverse order of removal after inspecting the key and keyway for damage or wear.

Breaker Point Removal and Installation

Remove the breaker cover. Care should be taken when removing the cover, to avoid damaging it. If the cover is bent or damaged, it should be replaced to insure a proper seal.

The breaker point gap on all models is 0.020 in. Check the points for contact and for signs of burning or pitting. Points that are set too wide will advance the spark timing and may cause kickback when starting. Points that are set too close will retard the spark timing and decrease engine power.

On models that have a separate condenser, the point set is removed by first removing the condenser and armature wires from the breaker point clip. Loosen the adjusting lock screw and remove the breaker point assembly.

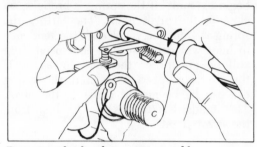

Removing the breaker point assembly

On models where the condenser is incorporated with the breaker points, loosen the screw which holds the post. The condenser/point assembly is removed by loosening the screw which holds the condenser clamp.

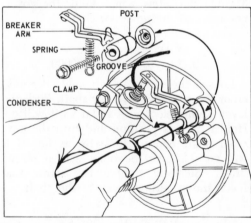

Removing the integral breaker point and condenser assembly

When installing a point set with the separate condenser, be sure that the small boss on the magneto plate enters the hole in the point bracket. Mount the point set to the magneto plate or the cylinder with a lock screw. Fasten the armature lead wire to the breaker points with the clip and screw. If these lead wires do not have terminals, the bare end of the wires can be inserted into the clip and the screw tightened to make a good connection. Do not let the ends of the wire touch either the point bracket or the magneto plate, or the ignition will be grounded.

To install the integral condenser/point

set, place the mounting post of the breaker arm into the recess in the cylinder so that the groove in the post fits the notch in the recess. Tighten the mounting screw securely. Use a ¼ in. wrench. Slip the open loop of the breaker arm spring through the two holes in the arm, then hook the closed loop of the spring over the small post protruding from the cylinder. Push the flat end of the breaker arm into the groove in the mounting post. This places tension on the spring and pulls the arm against the plunger. If the condenser post is threaded, attach the coil primary wire and the ground wire (if furnished) with the lock washer and nut. If the primary wire is fastened to the condenser with a spring fastener, compress the spring and slip the primary wire and ground wire into the hole in the condenser post. Release the spring. Lay the condenser in place and tighten the condenser clamp securely. Install the spring in the breaker arm.

Point Gap Adjustment

Turn the crankshaft until the points are open to the widest gap. When adjusting a

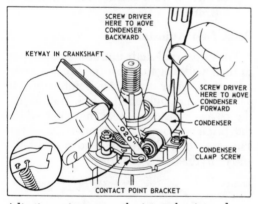

Adjusting point gap on the integral point and condenser assembly

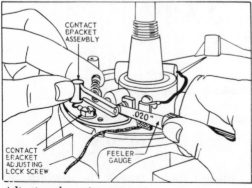

Adjusting the point gap

breaker point assembly with an integral condenser, move the condenser forward or backward with a screwdriver until the proper gap is obtained (0.020 in.). Point sets with a separate condenser are adjusted by moving the contact point bracket up and down after the lock screw has been loosened. The point gap is set to 0.020 in.

Breaker Point Plunger

If the breaker point plunger hole becomes excessively worn, oil will leak past the plunger and may get on the points, causing them to burn. To check the hole, loosen the breaker point mounting screw and move the breaker points out of the way. Remove the plunger. If the flat end of the #19055 plug gauge will enter the plunger hole for a distance of ¼ in. or more, the hole should be rebushed.

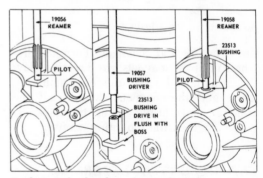

Replacing the breaker plunger bushing

To install the bushing, it is necessary that the breaker points, armature, and crankshaft be removed. Use a #19056 reamer to ream out the old plunger hole. This should be done by hand. The reamer must be in alignment with the plunger hole. Drive the bushing, #23513, into the hole until the upper end of the bushing is flush with the top of the boss. Remove all metal chips and dirt from the engine.

If the breaker point plunger is worn to a length of 0.870 in. or less, it should be replaced. Plungers must be inserted with the groove at the top or oil will enter the breaker box. Insert the plunger into the hole in the cylinder.

Armature Air Gap Adjustment

Set the air gap between the flywheel and the armature as follows: With the armature up as far as possible and just one screw tightened, slip the proper gauge between

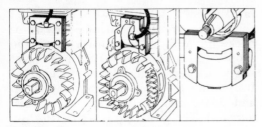

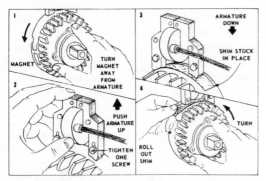

Variations in armature positioning

Adjustment of the armature gap

the armature and flywheel. Turn the fly-wheel until the magnets are directly below the armature. Loosen the one mounting screw and the magnets should pull the armature down firmly against the thickness gauge. Tighten the mounting screws.

FLYWHEEL TYPE—EXTERNAL BREAKER

Breaker Point Set Removal and Installation

Turn the crankshaft until the points open to their widest gap. This makes it easier to assemble and adjust the points later if the crankshaft is not removed. Remove the condenser and upper and lower mounting screws. Loosen the lock nut and back off the breaker point screw. Install the points in the reverse order of removal.

To avoid the possibility of oil leaking past the breaker point plunger or moisture entering the crankcase between the plunger and the bushing, a plunger seal is installed on the engine models using this type of ignition system. To install a new seal on the plunger, remove the breaker point as-sembly and condenser. Remove the retainer and eyelet, remove the old seal, and install the new one. Use extreme care when in-stalling the seal on the plunger to avoid damaging the seal. Replace the eyelet and

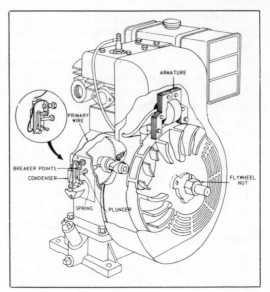

Flywheel magneto ignition with an external breaker assembly

retainer and replace the points and con-denser.

NOTE: *Apply a small amount of sealer to the threads of both mounting screws and the adjustment screw. The sealer prevents oil from leaking into the breaker point area.*

Point Gap Adjustment

Turn the crankshaft until the points open to their widest gap. Turn the breaker point adjusting screw until the points open to 0.020 in. and tighten the lock nut. When the cover is installed, seal the point where the primary wire passes under the cover. This area must be resealed to prevent the entry of dust and moisture.

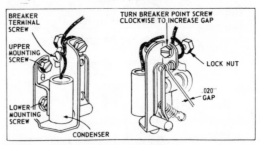

Breaker point gap adjustment

Armature Timing Adjustment

MODELS 193000, 200000, 230000, 243000, 300400, 320400

Using a puller, remove the flywheel. Set the point gap at 0.020 in. Position the flywheel on the crankshaft taper. Slip the key in place. Install the flywheel nut finger tight. Rotate the flywheel and the crankshaft clockwise until the breaker points are just opening. Use a timing light. When the points just start to open, the arrow on the flywheel should line up with the arrow on the armature bracket.

If the arrows do not match, slip off the flywheel without disturbing the position of the crankshaft. Slightly loosen the mounting screw which holds the armature bracket to the cylinder. Slip the flywheel back onto the crankshaft. Insert the flywheel key. Install the flywheel nut finger tight. Move the armature and bracket assembly to align the arrows. Slip off the flywheel and tighten the armature bracket bolts. Install the key and flywheel. Tighten the flywheel nut to 110 to 118 ft lbs on the 193000 and 200000 series. On all the rest, tighten to 138 to 150 ft lbs. Set the armature gap at 0.010 to 0.014 in.

MODELS 19D, 23D

With the points set at 0.020 in. and the flywheel key screw finger tight together with the flywheel nut, rotate the flywheel clockwise until the breaker points are just opening. The flywheel key drives the crankshaft while doing this. Using a timing light, rotate the flywheel slightly counterclockwise until the edge of the armature lines up with the edge of the flywheel insert. The crankshaft must not turn while doing this. Tighten the key screw and the flywheel nut. Set the armature air gap at 0.022 to 0.026 in.

Replacing the Breaker Plunger and Bushing

There are two types of bushings used: threaded and unthreaded.

THREADED

Remove the breaker cover and the condenser and breaker point assembly.

Place a thick $\frac{3}{8}$ in. inside diameter washer over the end of the bushing and screw on the $\frac{3}{8}$-24 nut. Tighten the nut to pull the bushing out of the hole. After the bushing has been moved about $\frac{1}{8}$ in., remove the nut and put on a second thick washer and repeat the procedure. A total stack of $\frac{3}{8}$ in. washers will be required to completely remove the bushing. Be sure

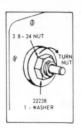
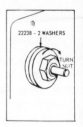

Removing a threaded plunger bushing

the plunger does not fall out of the bushing as it is removed.

Place the new plunger in the bushing with the large end of the plunger opposite the threads on the bushing. Screw the $\frac{3}{8}$-24 in. nut onto the threads to protect them and insert the bushing into the cylinder. Place a piece of tubing the same diameter as the nut and, using a hammer, drive the bushing into the cylinder until the square shoulder on the bushing is flush with the face of the cylinder. Check to be sure that the plunger operates freely.

UNTHREADED

Pull the plunger out as far as possible and use a pair of pliers to break the plunger off as close as possible to the bushing. Use a $\frac{1}{4}$-20 in. tap or a #93029 self threading screw to thread the hole in the bushing to a depth of about $\frac{1}{2}$–$\frac{5}{8}$ in. Use a $\frac{1}{4}$-20 x $\frac{1}{2}$ in. Hex. head screw and two spacer washers to pull the bushing out of the cylinder. The bushing will be free when it has been extracted $\frac{5}{16}$ in. Carefully remove the bushing and the remainder of the broken plunger. Do not allow the plunger or metal chips to drop into the crankcase.

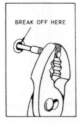

Removing an unthreaded plunger bushing

Correctly insert the new plunger into the new bushing. Insert the plunger and the bushing into the cylinder. Use a hammer and the old bushing to drive the new bushing into the cylinder until the new bushing is flush with the face of the cylinder. Make sure that the plunger operates freely.

MAGNA-MATIC IGNITION SYSTEM

Armature Air Gap

The armature air gap on engines equipped with Magna-Matic ignition system is fixed and can change only if wear occurs on the crankshaft journal and/or main bearing. Check for wear by inserting a ½ in. wide feeler gauge at several points between the rotor and armature. Minimum feeler gauge thickness is 0.004 in. Keep the feeler gauge away from the magnets on the rotor or you will have a false reading.

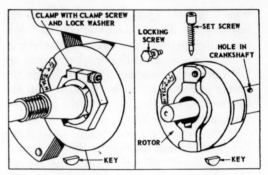

Removing the rotor

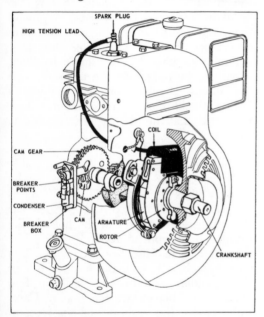

The Magna-Matic ignition system

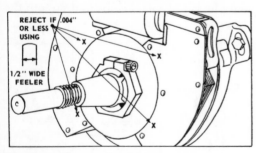

Measuring the armature gap

Rotor Removal and Installation

The rotor is held in place by a woodruff key and a clamp on later engines, and a woodruff key and set screw on older engines. The rotor clamp must always remain on the rotor, unless the rotor is in place on the crankshaft and within the armature, or a loss of magnetism will occur.

Loosen the socket head screw in the rotor clamp which will allow the clamp to loosen. It may be necessary to use a puller to remove the rotor from the crankshaft. On older models, loosen the small lock screw, then the set screw.

To install the set screw type rotor, place the woodruff key in the keyway on the crankshaft, then slide the rotor onto the crankshaft until the set screw hole in the rotor and the crankshaft are aligned. Be sure the key remains in place. Tighten the set screw securely, then tighten the lock screw to prevent the set screw from loosening. The lock screw is self-threading and the hole does not require tapping.

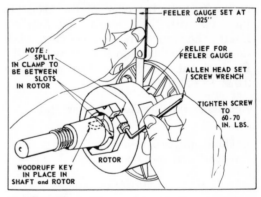

Installing the rotor

To install the clamp type rotor, place the woodruff key in place in the crankshaft and align the keyway in the rotor with the woodruff key. If necessary, use a short length of pipe and a hammer to drive the rotor onto the shaft until a 0.025 in. feeler gauge can be inserted between the rotor and the bearing support. The split in the clamp must be between the slots in the rotor. Tighten the clamp screws to 60 to 70 in. lbs.

Rotor Timing Adjustment

The rotor and armature are correctly timed at the factory and require timing only if the armature has been removed from the engine, or if the cam gear or crankshaft has been replaced.

If it is necessary to adjust the rotor, proceed as follows: with the point gap set at 0.020 in., turn the crankshaft in the normal direction of rotation until the breaker points close and just start to open. Use a timing light or insert a piece of tissue paper between the breaker points to determine when the points begin to open. With the three armature mounting screws slightly loose, rotate the armature until the arrow on the armature lines up with the arrow on the rotor. Align with the corresponding number of engine models, for example, on Model 9, align with #9. Retighten the armature mounting screws.

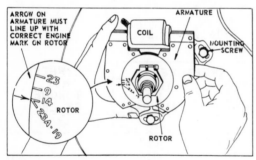

Adjustment of the timing

Coil and/or Armature Replacement

Usually the coil and armature are not separated, but left assembled for convenience. However, if one or both need replacement, proceed as follows: the coil primary wire and the coil ground wire must be unfastened. Pry out the clips that hold the coil and coil core to the armature. The coil core is a slip fit in the coil and can be pushed out of the coil.

To reassemble, push the coil core into the coil with the rounded side toward the ignition cable. Place the coil and core on the armature with the coil retainer between the coil and the armature and with the rounded side toward the coil. Hook the lower end of the clips into the armature, then press the upper end onto the coil core.

Fasten the coil ground wire (bare double wires) to the armature support. Next, place the assembly against the cylinder and around the rotor and bearing support. Insert the three mounting screws together with the washer and lockwasher into the three long oval holes in the armature. Tighten them enough to hold the armature in place but loose enough so the armature can be moved for adjustment of the timing. Attach the primary wires from the coil and the breaker points to the terminal at the upper side of the backing plate. This terminal is insulated from the backing plate. Push the ignition cable through the louvered hole at the left side of the backing plate.

NOTE: *On Model 9 engines, knot the ignition cable before inserting it through the backing plate. Be sure all wires are clear of the flywheel.*

Breaker Point Removal and Installation

Turn the crankshaft until the points open to the widest gap. This makes it easier to assemble and adjust the points later if the crankshaft is not removed. With the terminal screw out, remove the spring screw. Loosen the breaker shaft nut until the nut is flush with the end of the shaft. Tap the nut to free the breaker arm from the tapered end of the breaker shaft. Remove the nut, lockwasher, and breaker arm. Remove the breaker plate screw, breaker plate, pivot, insulating plate, and eccentric. Pry out the breaker shaft seal with a sharp pointed tool.

To install the breaker points, press in the new oil seal with the metal side out. Put the new breaker plate on the top of the insulating plate, making sure that the detent in the breaker plate engages the hole in the insulating plate. Fasten the breaker plate screw enough to put a light tension on the plate. Adjust the eccentric so that the left edge of the insulating plate is parallel to the edge of the box and tighten the screw. This locates the breaker plate so that the proper gap adjustments may be made. Turn the breaker shaft clockwise as far as possible and hold it in this position. Place the new breaker points on the shaft, then the lockwasher, and tighten the nut down on the lockwasher. Replace the spring screw and terminal screw.

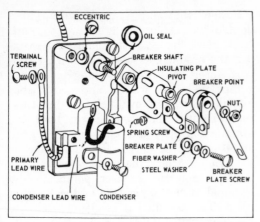

The breaker point assembly

the hole at the lower left corner of the breaker box. See that the primary wire rests in the groove at the top end of the box, then tighten the two mounting screws to hold the box in place.

Breaker Box Removal and Installation

Remove the two mounting screws, then remove the breaker box, turning it slightly to clear the arm at the inner end of the breaker shaft. The breaker points need not be removed to remove the breaker box.

To install, pull the primary wire through

Breaker Shaft Removal and Installation

The breaker shaft can be removed, after the breaker points are removed, by turning the shaft one half turn to clear the retaining spur at the inside of the breaker box.

Install by inserting the breaker shaft with the arm upward so the arm will clear the retainer boss. Push the shaft all the way in, then turn the arm downward.

Breaker Point Adjustment

To adjust the breaker points, turn the crankshaft until the breaker points open to the widest gap. Loosen the breaker point plate screw slightly. Rotate the eccentric to obtain a point gap of 0.020 in. Tighten the breaker plate screw.

1. Spark plug gap: .030
2. Condenser capacity: .18 to .24 M.F.D.
3. Contact point gap: .020

BASIC MODEL SERIES	ARMATURE TWO LEG AIR GAP	ARMATURE THREE LEG AIR GAP	FLYWHEEL NUT TORQUE FT. LBS.
ALUMINUM CYLINDER			
6B, 60000, 8B	.006 .010	.012 .016	57
80000, 82000, 92000	.006 .010	.012 .016	57
100000, 130000	.010 .014	.012 .016	60
140000, 170000, 190000	.010 .014	.016 .019	67
CAST IRON CYLINDER			
5, 6, N, 8		.012 .016	57
9			60
14			67
19, 190000, 200000	.010 .014	.022 .026	114
23, 230000	.010 .014	.022 .026	144
243400, 300000, 320000	.010 .014		144

Tune-up specifications

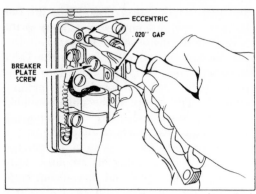

Adjusting the breaker point gap

Engine Mechanical

CYLINDER HEAD

Compression Checking

You can check the compression in any Briggs and Stratton engine by performing the following simple procedure: spin the flywheel counterclockwise (flywheel side) against the compression stroke. A sharp rebound indicates that there is satisfactory compression. A slight or no rebound indicates poor compression.

It has been determined that this test is an acurate indication of compression and is recommended by Briggs and Stratton. Briggs and Stratton does not supply compression pressures.

Loss of compression will usually be the result of one or a combination of the following:

1. The cylinder head gasket is blown or leaking.

2. The valves are sticking or not seating properly.

3. The piston rings are not sealing, which would also cause the engine to consume an excessive amount of oil.

Carbon deposits in the combustion chamber should be removed every 100 or 200 hours of use (more often when run at a steady load), or whenever the cylinder head is removed.

Removal and Installation

Always note the position of the different cylinder head screws so that they can be properly reinstalled. If a screw is used in the wrong position, it may be too short and not engage enough threads. If it is too long, it may bottom on a fin, either breaking the fin, or leaving the cylinder head loose.

Remove the cylinder screws and then the cylinder head. Be sure to remove the gasket and all remaining gasket material from the cylinder head and the block.

Assemble the cylinder head with a new gasket, cylinder head shield, screws, and washers in their proper places. Graphite grease should be used on aluminum cylinder head screws.

Do not use a sealer of any kind on the head gasket. Tighten the screws down evenly by hand. Use a torque wrench and tighten the head bolts in the correct sequence.

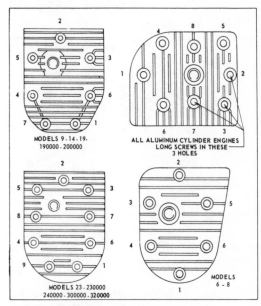

Cylinder head bolt tightening sequences

BASIC MODEL SERIES	IN. LBS. TORQUE
ALUMINUM CYLINDER	
6B, 60000, 8B, 80000 82000, 92000 100000, 130000	140
140000, 170000, 190000	165
CAST IRON CYLINDER	
5, 6, N, 8, 9	140
14	165
19, 190000, 200000, 23, 230000, 240000, 300000, 320000	190

Cylinder head bolt torque specifications

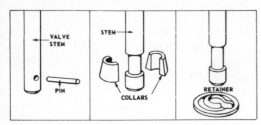

The three types of valve spring retainers

VALVES

Removal and Installation

Using a valve spring compressor, place one of the jaws over the valve spring and the other underneath, between the spring and the valve chamber. This positioning of the valve spring compressor is for valves that have either pin or collar type retainers. Tighten the jaws to compress the spring. Remove the collars or pin and lift out the valve. Pull out the compressor and the spring.

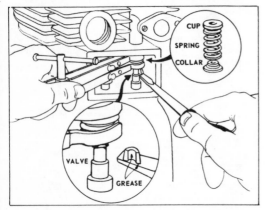

Removing the valve springs with the help of a valve spring compressor

To remove valves with ring type retainers, position the compressor with the upper jaw over the top of the valve chamber and the lower jaw between the spring and the retainer. Compress the spring, remove the retainer, and pull out the valve. Remove the compressor and spring.

Before installing the valves, check the thickness of the valve springs. Some engines use the same spring for the intake and exhaust side, while others use a heavier spring on the exhaust side. Compare the springs before installing them.

If the retainers are held by a pin or collars, place the valve spring and retainer

and cup (Models 9-14-19-20-23-24-32) into the valve spring compressor. Compress the spring until it is solid. Insert the compressed spring and retainer into the valve chamber. Then drop the valve into place, pushing the stem through the retainer. Hold the spring up in the chamber, hold the valve down, and insert the retainer pin with needle nose pliers or place the collars in the groove in the valve stem. Loosen the spring until the retainer fits around the pin or collars, then pull out the spring compressor. Be sure the pin or collars are in place.

To install valves with ring type retainers, compress the retainer and spring with the compressor. The large diameter of the retainer should be toward the front of the valve chamber. Insert the compressed spring and retainer into the valve chamber. Drop the valve stem through the larger area of the retainer slot and move the compressor so as to center the small area of the valve retainer slot onto the valve stem shoulder. Release the spring tension and remove the compressor.

Refacing Valve and Seats

Faces on valves and valve seats should be resurfaced with a valve grinder or cutter to an angle of 45°.

NOTE: *Some engines have a 30° intake valve and seat.*

The valve and seat should then be lapped with a fine lapping compound to remove the grinding marks and ensure a good seat. The valve seat width should be $\frac{3}{64}$–$\frac{1}{16}$ in. If the seat is wider, a narrowing stone or cutter should be used. If either the seat or valve is badly burned, it should be replaced. Replace the valve if the edge thickness (margin) is less than $\frac{1}{64}$ in. after it has been resurfaced.

MODEL SERIES	INTAKE		EXHAUST	
	MAX.	MIN.	MAX.	MIN.
ALUMINUM CYLINDER				
6B, 60000, 8B, 80000	.007	.005	.011	.009
82000, 92000, 100000	.007	.005	.011	.009
130000, 140000, 170000, 190000	.007	.005	.011	.009
CAST IRON CYLINDER				
5, 6, 8, N, 9, 14, 19 190000, 200000	.009	.007	.016	.014
23, 230000, 240000 300000, 320000	.009	.007	.019	.017

Valve clearances

Check and Adjust Tappet Clearance

Insert the valves in their respective positions in the cylinder. Turn the crankshaft until one of the valves is at its highest position. Turn the crankshaft one revolution. Check the clearance with a feeler gauge. Repeat for the other valve. Grind off the end of the valve stem if necessary to obtain proper clearance.

NOTE: *Check the valve tappet clearance with the engine cold.*

VALVE SEAT INSERTS

Cast iron cylinder engines are equipped with an exhaust valve insert which can be removed and replaced with a new insert. The intake side must be counterbored to allow the installation of an intake valve seat insert. Aluminum alloy cylinder models are equipped with inserts on both the exhaust and intake valves.

Removal and Installation

Valve seat inserts are removed with a special puller.

NOTE: *On Aluminum alloy cylinder models, it may be necessary to grind the puller nut until the edge is ⅟₃₂ in. thick in order to get the puller nut under the valve insert.*

When installing the valve seat insert, make sure that the side with the chamfered outer edge goes down into the cylinder. Install the seat insert and drive it into place with a driver. The seat should then be ground lightly and the valves and seats lapped lightly with grinding compound.

NOTE: *Aluminum alloy cylinder models*

use the old insert as a spacer between the driver and the new insert. Drive in the new insert until it bottoms. The top of the insert will be slightly below the cylinder head gasket surface. Peen around the insert using a punch and hammer.

NOTE: *The intake valve seat on cast iron cylinder models has to be counterbored before installing the new valve seat insert.*

PISTONS, PISTON RINGS, AND CONNECTING RODS

Removal

To remove the piston and connecting rod from the engine, bend down the connecting rod lock. Remove the connecting rod cap. Remove any carbon or ridge at the top of the cylinder bore. This will prevent breaking the rings. Push the piston and rod out of the top of the cylinder.

Pistons used in sleeve bore, aluminum alloy engines are marked with an "L" on top of the piston. These pistons are tin plated and use an expander with the oil ring. This piston assembly is not interchangeable with the piston used in the aluminum bore engines (Kool bore).

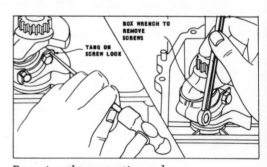

Removing the connecting rod cap

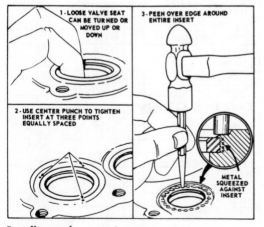

Installing valve seat inserts

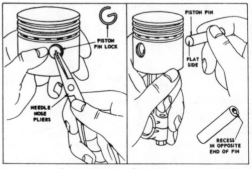

Removing the wrist pin and connecting rod from the piston

Pistons used in aluminum bore (Kool bore) engines are not marked on the top.

To remove the connecting rod from the piston, remove the piston pin lock with thin nose pliers. One end of the pin is drilled to facilitate removal of the lock.

Remove the rings one at a time, slipping them over the ring lands. Use a ring expander to remove the rings.

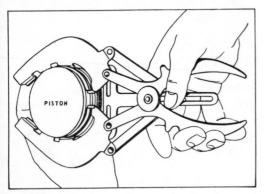

Replacing the piston rings

Inspection

Check the piston ring fit. Use a feeler gauge to check the side clearance of the top ring. Make sure that you remove all carbon from the top ring groove. Use a new piston ring to check the side clearance. If the cylinder is to be resized, there is no reason to check the piston, since a new oversized piston assembly will be installed. If the side clearance is more than 0.007 in., the piston is excessively worn and should be replaced.

Check the piston ring end gap by cleaning all carbon from the ends of the rings and inserting them one at a time 1 in. down into the cylinder. Check the end gap with a feeler gauge. If the gap is larger than recommended, the ring should be replaced.

NOTE: *When checking the ring gap, do not deglaze the cylinder walls by installing piston rings in aluminum cylinder engines.*

Chrome ring sets are available for all current aluminum and cast iron cylinder models. No honing or deglazing is required. The cylinder bore can be a maximum of 0.005 in. oversize when using chrome rings.

If the crankpin bearing in the rod is scored, the rod must be replaced. 0.005 in. oversize piston pins are available in case the connecting rod and piston are worn at the piston pin bearing. If, however, the crankpin bearing in the connecting rod is worn, the rod should be replaced. Do not attempt to file or fit the rod.

If the piston pin is worn 0.0005 in. out of round or below the rejection sizes, it should be replaced.

Installation

The piston pin is a push fit into both the piston and the connecting rod. On models using a solid piston pin, one end is flat and the other end is recessed. Other

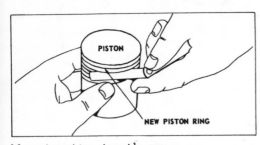

Measuring piston ring side gap

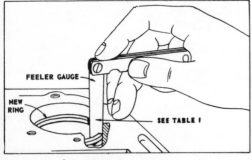

Measuring the piston ring gap

BASIC MODEL SERIES	CRANK PIN BEARING	PISTON PIN BEARING
ALUMINUM CYLINDER		
6B, 60000	.876	.492
8B, 80000	1.0013	.492
82000, 92000	1.0013	.492
100000	1.0013	.555
130000	1.0013	.492
140000, 170000	1.0949	.674
190000	1.1265	.674
CAST IRON CYLINDER		
5	.7524	.492
6, 8. N	.751	.492
9	.876	.5633
14, 19, 190000	1.0007	.6735
200000	1.1265	.6735
23, 230000	1.189	.736
240000	1.314	.6735
300000, 320000	1.314	.8015

Connecting rod bearing specifications

BASIC MODEL SERIES	COMP. RING	OIL RING
ALUMINUM CYLINDER		
6B, 60000, 8B, 80000		
82000, 92000	.035	.045
100000, 130000		
140000, 170000, 190000		
CAST IRON CYLINDER		
5, 6, 8, N, 9		
14, 19, 190000		
200000, 23	.030	.035
230000, 240000		
300000, 320000		

Piston ring gap specifications

BASIC MODEL SERIES	PISTON PIN	PIN BORE
ALUMINUM CYLINDER		
6B, 60000	.489	.491
8B, 80000	.489	.491
82000, 92000	.489	.491
100000	.552	.554
130000	.489	.491
140000, 170000, 190000	.671	.671
CAST IRON CYLINDER		
5, 6, 8, N	.489	.491
9	.561	.563
14, 19, 190000	.671	.673
200000	.671	.673
23, 230000	.734	.736
240000	.671	.673
300000, 320000	.799	.801

Wrist pin specifications

models use a hollow piston pin. Place a pin lock in the groove at one side of the piston. From the opposite side of the piston, insert the piston pin, flat end first for solid pins; with hollow pins, insert either end first until it stops against the pin lock. Use thin nose pliers to assemble the pin lock in the

recessed end of the piston. Be sure the locks are firmly set in the groove.

Install the rings on the pistons, using a piston ring expander. Make sure that they are installed in the proper position. The scraper groove on the center compression ring should always be down toward the piston skirt. Be sure the oil return holes are clean and all carbon is removed from the grooves.

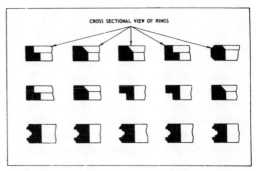

Cross-sectional views and positioning of the various types of piston rings used in Briggs and Stratton engines

NOTE: *Install the expander under the oil ring in sleeve bore aluminum alloy engines.*

Oil the rings and the piston skirt, then compress the rings with a ring compressor. Turn the piston and compressor upside down on the bench and push downward so the piston head and the edge of the compressor band are even, all the while tightening the compressor. Draw the compressor up tight to fully compress the rings, then loosen the compressor very slightly.

CAUTION: *Do not attempt to install the piston and ring assembly without using a ring compressor.*

Place the connecting rod and piston as-

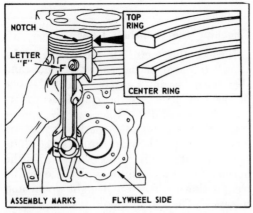

Assembling the piston and connecting rod assembly

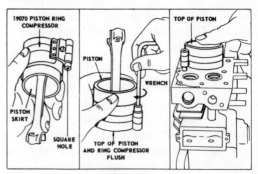

Installing the piston and connecting rod assembly into the cylinder block

sembly, with the rings compressed, into the cylinder bore. Push the piston and rod down into the cylinder. Oil the crankpin of the crankshaft. Pull the connecting rod against the crankpin and assemble the rod cap so the assembly marks align.

NOTE: *Some rods do not have assembly marks, as the rod and cap will fit together only in one position. Use care to ensure proper installation.*

Assemble the cap screw and screw locks with the oil dippers (if used). Tighten the cap screws to the crankshaft and turn the crankshaft two revolutions to be sure the rod is correctly installed. If the rod strikes the camshaft, the connecting rod has been installed wrong or the cam gear is out of time. If the crankshaft operates freely, bend the cap screw locks against the screw heads.

Securely tighten the rod screws. After tightening the rod screws, the rod should be able to move sideways on the crankpin of the shaft. Use a torque wrench to prevent loose or overtightened cap screws which would result in breakage and/or rod scoring.

CRANKSHAFT AND CAMSHAFT GEAR

Removal

ALUMINUM CYLINDER ENGINES

To remove the crankshaft from aluminum alloy engines, remove any rust or burrs from the power take-off end of the crankshaft. Remove the crankcase cover or sump. If the sump or cover sticks, tap it lightly with a soft hammer on alternate sides near the dowel. Turn the crankshaft to align the crankshaft and camshaft timing marks, lift out the cam gear, then remove the crankshaft. On models that have ball bearings on the crankshaft, the crankshaft and the camshaft must be removed together.

CAST IRON CYLINDER MODELS

To remove the crankshaft from cast iron models (9-14-19-190000-200000-23-230000-240000-300000-320000), remove the crankcase cover. Revolve the crankshaft until the crankpin is pointing upward toward the breather at the rear of the engine (approximately a 45° angle). Pull the crankshaft

out from the drive side, twisting it slightly if necessary. On models with ball bearings on the crankshaft, both the crankcase cover and bearing support should be removed.

On cast iron models with ball bearings on the drive side, first remove the magneto. Drive out the camshaft. Push the camshaft forward into the recess at the front of the engine. Then draw the crankshaft from the magneto side of the engine. Double thrust engines have cap screws inside the crankcase which hold the bearing in place. These must be removed before the crankshaft can be removed.

To remove the camshaft from all cast iron models, except the 300400 and 320400, use a long punch to drive the camshaft out toward the magneto side. Save the plug. Do not burr or peen the end of the shaft while driving it out. Hold the camshaft while removing the punch, so it will not drop and become damaged.

Checking the Crankshaft

Discard the crankshaft if it is worn beyond the allowable limit. Check the keyways for wear and make sure they are not spread. Remove all burrs from the keyway to prevent scratching the bearing. Check the three bearing journals, drive end, crankpin, and magneto end, for size and any wear or damage. Check the cam gear teeth for wear. They should not be worn at all. Check the threads at the magneto end for damage. Make sure that the crankshaft is straight.

NOTE: *There are 0.020 in. undersize connecting rods available for use on reground crankpin bearings.*

Checking the Camshaft Gear

Inspect the teeth for wear and nicks. Check the size of the camshaft and camshaft gear bearing journals. Check the size of the cam lobes. If the cam is worn beyond tolerance, discard it.

Check the automatic spark advance on models equipped with the Magna-Matic ignition system. Place the cam gear in the normal operating position with the movable weight down. Press the weight down and release it. The spring should lift the weight. If not, the spring is stretched or the weight is binding.

BASIC MODEL SERIES	CAM GEAR OR SHAFT JOURNALS	CAM LOBE
ALUMINUM CYLINDER		
6B, 60000	.4985	.883
8B, 80000*	.4985	.883
82000, 92000	.4985	.883
100000, 130000	.4985	.950
140000, 170000, 190000	.4985	.977
CAST IRON CYLINDER		
5, 6, 8, N	.3719	.875
9	.3719	1.124
14, 19, 190000	.4968	1.115
200000	.4968	1.115
23, 230000	.4968	1.184
240000	.4968	1.184
300000	#	1.184
320000	#	1.215

* Auxiliary drive models P.T.O. .751
\# Mag. side − .8105   P.T.O. side − .6145
Camshaft specifications

BASIC MODEL SERIES	P.T.O. JOURNAL	MAG. JOURNAL	CRANKPIN
ALUMINUM CYLINDER			
6B, 60000	.8726	.8726	.8697
8B, 80000*	.8726	.8726	.9963
82000, 92000*	.8726	.8726	.9963
100000, 130000	.9976	.8726	.9963
140000, 170000	1.1790	.9975#	1.090
190000	1.1790	.9975#	1.1219
CAST IRON CYLINDER			
5, 6, 8, N	.8726	.8726	.7433
9	.9832	.9832	.8726
14, 19, 190000	1.1790	1.1790	.9964
200000	1.1790	1.1790	1.1219
23, 230000	1.3759	1.3759	1.1844
240000	Ball	Ball	1.3094
300000, 320000	Ball	Ball	1.3094

* Auxiliary drive models P.T.O. Bearing
  Reject size − 1.003

\# Synchro Balanced Magneto Bearing
  Reject size − 1.179

Crankshaft specifications

Removal and Installation of the Ball Bearings

The ball bearings are pressed onto the crankshaft. If either the bearing or the crankshaft is to be removed, use an arbor press to remove them.

To install, heat the bearing in hot oil (325° F maximum). Don't let the bearing rest on the bottom of the pan in which it is heated. Place the crankshaft in a vise with the bearing side up. When the bearing is quite hot, it will slip fit onto the bearing journal. Grasp the bearing, with the shield down, and thrust it down onto the crank-

shaft. The bearing will tighten on the shaft while cooling. Do not quench the bearing (throw water on it to cool it).

Installation

ALUMINUM ALLOY ENGINES— PLAIN BEARING

In aluminum alloy engines, the tappets are inserted first, the crankshaft next, and then the cam gear. When inserting the cam gear, turn the crankshaft and the cam gear so that the timing marks on the gears align.

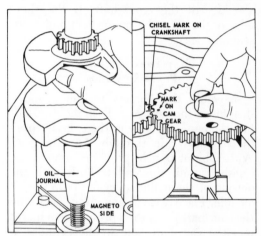

Alignment of the camshaft and crankshaft timing marks

ALUMINUM ALLOY ENGINES— BALL BEARING

On crankshafts with ball bearings, the gear teeth are not visible for alignment of the timing marks; therefore, the timing mark is on the counterweight. On ball bearing equipped engines, the tappets are installed first. The crankshaft and the cam gear must be inserted together and their timing marks aligned.

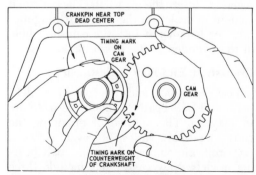

Alignment of the camshaft and crankshaft timing marks on engines equipped with ball bearings

CRANKCASE COVER
AND CRANKSHAFT

Installation

MODELS 100900 AND 130900

On these models, install the governor slinger onto the cam gear with the spring washer.

To protect the oil seal while assembling the crankcase cover, put oil or grease on the sealing edge of the oil seal. Wrap a piece of thin cardboard around the crankshaft so the seal will slide easily over the shoulder of the crankshaft. If the sharp edge of the oil seal is cut or bent under, the seal may leak.

CAST IRON ENGINES— PLAIN BEARINGS

Assemble the tappets to the cylinder, then insert the cam gear. Push the camshaft into the camshaft hole in the cylinder from the flywheel side through the cam gear. With a blunt punch, press or hammer the camshaft until the end is flush with the outside of the cylinder on the power take-off side. Place a small amount of sealer on the camshaft plug, then press or hammer it into the camshaft hole in the cylinder at the flywheel side. Install the crankshaft so the timing marks on the teeth and on the cam gear align.

CAST IRON—BALL BEARINGS

Assemble the tappets, then insert the cam gear into the cylinder, pushing the cam gear forward into the recess in front of the cylinder. Insert the crankshaft into the cylinder. Turn the camshaft and crankshaft until the timing marks align, then push the cam gear back until it engages the gear on the crankshaft with the timing marks together. Insert the camshaft. Place a small amount of sealer on the camshaft plug and press or hammer it into the camshaft hole in the cylinder at the flywheel side.

Crankshaft End-Play Adjustment

The crankshaft end-play on all models, plain and ball bearing, should be 0.002 in. to 0.008 in. The method of obtaining the correct end-play varies, however, between cast iron, aluminum, plain, and ball bearing models. New gasket sets include three crankcase cover or bearing support gaskets, 0.005 in., 0.009 in., and 0.015 in. thick.

The end-play of the crankshaft may be checked by assembling a dial indicator on the crankshaft with the pointer against the crankcase. Move the crankshaft in and out. The indicator will show the end-play. Another way to measure the end-play is to assemble a pulley to the crankshaft and measure the end-play with a feeler gauge. Place the feeler gauge between the crankshaft thrust face and the bearing support. The feeler gauge method of measuring crankshaft end-play can only be used on cast iron plain bearing engines with removable bases.

The end-play should be 0.002 in. to 0.008 in. with one 0.015 in. gasket in place. If the end-play is less than 0.002 in., which would be the case if a new crankcase or sump cover is used, additional gaskets of 0.005 in., 0.009 in., or 0.015 in. may be added in various combinations to obtain the proper end-play.

ALUMINUM ENGINES ONLY

If the end-play is more than 0.008 in. with one 0.015 in. gasket in place, a thrust washer is available to be placed on the crankshaft power take-off end, between the gear and crankcase cover or sump on plain bearing engines. On ball bearing equipped aluminum engines, the thrust washer is added to the magneto end of the crankshaft instead of the power take-off end.

NOTE: *Aluminum engines never use less than the 0.015 in. gasket.*

CYLINDERS

Inspection

Always inspect the cylinder after the engine has been disassembled. Visual inspection will show if there are any cracks, stripped bolt holes, broken fins, or if the cylinder wall is scored. Use an inside micrometer or telescoping gauge and micrometer to measure the size of the cylinder bore. Measure at right angles.

If the cylinder bore is more than 0.003 in. oversize, or 0.0015 in. out of round on lightweight (aluminum) cylinders, the cylinder must be resized (rebored).

NOTE: *Do not deglaze the cylinder walls when installing piston rings in aluminum cylinder engines. Also be aware that there are chrome ring sets available for most engines. These are used to control oil*

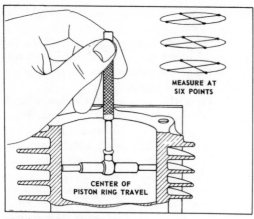

MEASURE AT
SIX POINTS

CENTER OF
PISTON RING TRAVEL

Checking the cylinder bore

pumping in bores worn to 0.005 in. over standard and do not require honing or glaze breaking to seat.

Resizing

Always resize to exactly 0.010 in., 0.020 in., or 0.030 in. over standard size. If this is done accurately, the stock oversize rings and pistons will fit perfectly and proper clearances will be maintained. Cylinders, either cast iron or lightweight, can be quickly resized with a good hone. Use the stones and lubrication recommended by the hone manufacturer to produce the correct cylinder wall finish for the various engine models.

If a boring bar is used, a hone must be used after the boring operation to produce the proper cylinder wall finish. Honing can be done with a portable electric drill, but it is easier to use a drill press.

1. Clean the cylinder at top and bottom to remove all burrs and pieces of base and head gaskets.

2. Fasten the cylinder to a heavy iron plate. Some cylinders require shims. Use a level to align the drill press spindle with the bore.

3. Oil the surface of the drill press table liberally. Set the iron plate and the cylinder on the drill press table. Do not anchor the cylinder to the drill press table. If you are using a portable drill, set the plate and the cylinder on the floor.

4. Place the hone driveshaft in the chuck of the drill.

5. Slip the hone into the cylinder. Connect the driveshaft to the hone and set the stop on the drill press so the hone can only extend ¾ in. to 1 in. from the top or bottom of the cylinder. If you are using a portable drill, cut a piece of wood to place in the cylinder as a stop for the hone.

6. Place the hone in the middle of the cylinder bore. Tighten the adjusting knob with your finger or a small screwdriver until the stones fit snugly against the cylinder wall. Do not force the stones against the cylinder wall. The hone should operate at a speed of 300–700 rpm. Lubricate the hone as recommended by the manufacturer.

NOTE: *Be sure that the cylinder and the hone are centered and aligned with the driveshaft and the drill spindle.*

7. Start the drill and, as the hone spins, move it up and down at the lower end of the cylinder. The cylinder is not worn at the bottom but is round so it will act to guide the hone and straighten the cylinder bore. As the bottom of the cylinder increases in diameter, gradually increase your strokes until the hone travels the full length of the bore.

NOTE: *Do not extend the hone more than ¾ in. to 1 in. past either end of the cylinder bore.*

8. As the cutting tension decreases, stop the hone and tighten the adjusting knob. Check the cylinder bore frequently with an accurate micrometer. Hone 0.0005 in. oversize to allow for shrinkage when the cylinder cools.

9. When the cylinder is within 0.0015 in. of the desired size, change from the rough stone to a finishing stone.

The finished resized cylinder should have a cross-hatched appearance. Proper stones, lubrication, and spindle speed along with rapid movement of the hone within the cylinder during the last few strokes, will produce this finish. Cross-hatching provides proper lubrication and ring break-in.

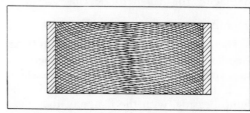

Cross hatch pattern after honing

NOTE: *It is EXTREMELY important that the cylinder be thoroughly cleaned after honing to eliminate ALL grit. Wash the cylinder carefully in a solvent such as*

BASIC ENGINE MODEL OR SERIES	STD. BORE SIZE DIAMETER	
ALUMINUM CYLINDER	MAX.	MIN.
6B		
60000 before Ser. #5810060	2.3125	2.3115
60000 after Ser. #5810030	2.375	2.374
8B, 80000, 82000	2.375	2.374
92000	2.5625	2.5615
100000	2.500	2.499
130000	2.5625	2.5615
140000	2.750	2.749
170000, 190000	3.000	2.999
CAST IRON CYLINDER		
5, 6, 5S, N	2.000	1.999
8	2.250	2.249
9	2.250	2.249
14	2.623	2.624
19, 23, 190000, 200000	3.000	2.999
230000	3.000	2.999
243400	3.0625	3.0615
300000	3.4375	3.4365
320000	3.5625	3.5615

Cylinder bore specifications

kerosene. The cylinder bore should be cleaned with a brush, soap, and water.

BEARINGS

Inspection

PLAIN TYPE

Bearings should be replaced if they are scored or if a plug gauge will enter. Try the gauge at several points in the bearing.

Replacing Plain Bearings

MODELS 9-14-19-20-23

The crankcase cover bearing support should be replaced if the bearing is worn or scored.

Replacing the Magneto Bearing

ALUMINUM CYLINDER ENGINES

There are no removable bearings in these engines. The cylinder must be reamed out so a replacement bushing can be installed.

1. Place a pilot guide bushing in the sump bearing, with the flange of the guide bushing toward the inside of the sump.

2. Assemble the sump on the cylinder. Make sure that the pilot guide bushing does not fall out of place.

3. Place the guide bushing into the oil seal recess in the cylinder. This guide bushing will center the counterbore reamer even

though the oil bearing surface might be badly worn.

4. Place the counterbore reamer on the pilot and insert them into the cylinder until the tip of the pilot enters the pilot guide bushing in the sump.

5. Turn the reamer clockwise with a steady, even pressure until it is completely through the bearing. Lubricate the reamer with kerosene or any other suitable solvent.

NOTE: *Counterbore reaming may be performed without any lubrication. However, clean off shavings because aluminum material builds up on the reamer flutes causing eventual damage to the reamer and an oversize counterbore.*

6. Remove the sump and pull the reamer out without backing it through the bearing. Clean out the remaining chips. Remove the guide bushing from the oil seal recess.

7. Hold the new bushing against the outer end of the reamed out bearing, with the notch in the bushing aligned with the notch in the cylinder. Note the position of the split in the bushing. At a point in the outer edge of the reamed out bearing opposite to the split in the bushing, make a notch in the cylinder hub at a 45° angle to the bearing surface. Use a chisel or a screwdriver and hammer.

8. Press in the new bushing, being careful to align the oil notches with the driver and the support until the outer end of the bushing is flush with the end of the reamed cylinder hub.

9. With a blunt chisel or screwdriver, drive a portion of the bushing into the notch previously made in the cylinder. This is called staking and is done to prevent the bushing from turning.

10. Reassemble the sump to the cylinder with the pilot guide bushing in the sump bearing.

11. Place a finishing reamer on the pilot and insert the pilot into the cylinder bearing until the tip of the pilot enters the pilot guide bushings in the sump bearing.

12. Lubricate the reamer with kerosene, fuel oil, or other suitable solvent, then ream the bushing, turning the reamer clockwise with a steady even pressure until the reamer is completely through the bearing. Improper lubricants will produce a rough bearing surface.

13. Remove the sump, reamer, and the pilot guide bushing. Clean out all reaming chips.

BASIC ENGINE MODEL SERIES	PTO BEARING	BEARING MAGNETO
ALUMINUM CYLINDER		
6B, 8B *	.878	.878
60000, 80000 *	.878	.878
82000, 92000 *	.878	.878
100000, 130000	1.003	.878
140000, 170000	1.185	1.0036#
190000	1.185	1.0036#
CAST IRON CYLINDER		
5, 6 N	.878	.878
9	.9875	.9875
14	1.185	1.185
19, 190000, 200000	1.185	1.185
23, 230000	1.382	1.382
240000, 300000	Ball	Ball
320000	Ball	Ball

Crankshaft bearing specifications

Lubrication

The primary purpose of oil is, of course, lubrication. However oil performs three other very important functions as well: cooling, cleaning, and sealing. Oil absorbs and dissipates heat created by combustion and friction. Oil cleans by trapping and holding dirt and by-products of combustion. This dirt is held in suspension by the oil until it is drained. Oil seals the combustion chamber by coating the rings, thus helping increase and maintain compression.

Briggs and Stratton engines are lubricated with a gear driven splash oil slinger or a connecting rod dipper.

Oil Changing and Checking

Change the oil after the first 5 hours of operation. Thereafter change the oil every 25 hours of operation; more often if the engine is operated under dusty conditions.

1. To change the oil, remove the drain plug and drain the oil while the engine is still warm.

2. Replace the drain plug.

3. Remove the oil filler plug or cap and refill with new oil of the proper grade and viscosity.

4. Replace the oil filler plug or cap. NOTE: *Do not use any type of special additives.*

5. Check the oil level regularly, at least after every 5 hours of operation.

If the unit is equipped with a gear reduction case, check the oil level in the case every time you check the engine oil. To check the oil level of the gear reduction, remove the oil level plug. If oil runs out, the level is satisfactory. If not, add oil through the oil fill hole until oil does run out of the oil level hole. Replace both the oil fill plug and the oil level plug.

In an engine where the gear reduction is lubricated with the same oil as the crankcase, change the oil every time the crankcase oil is changed. If the gear reduction is separate from the engine crankcase, the oil has to be changed only after every 100 hours of operation.

EXTENDED OIL FILLER TUBES AND DIPSTICKS

When installing the extended oil fill and dipstick assembly, the tube must be installed so the O-ring seal is firmly compressed. To do so, push the tube downward toward the sump, then tighten the blower housing screw, which is used to secure the tube and bracket. When the dipstick assembly is fully depressed, it seals the upper end of the tube.

A leak at the seal between the tube and the sump, or at the seal at the upper end of the dipstick can result in a loss of crankcase vacuum, and a discharge of smoke through the exhaust system.

BREATHERS

The function of the breather is to maintain a vacuum in the crankcase. The breather has a fiber disc valve which limits the direction of air flow caused by the piston moving back and forth in the cylinder. Air can flow out of the crankcase, but the one-way valve blocks the return flow, thus maintaining a vacuum in the crankcase. A partial vacuum must be maintained in the crankcase to prevent oil from being forced out of the engine at the piston rings, oil seals, breaker plunger, and gaskets.

Inspection of the Breather

If the fiber disc valve is stuck or binding, the breather cannot function properly and must be replaced. A 0.045 in. wire gauge should not enter the space between the fiber disc valve and the body. Use a spark plug wire gauge to check the valve. The fiber disc valve is held in place by an

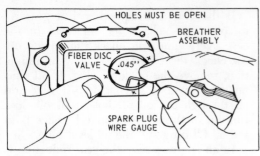

Checking the breather assembly

BASIC MODEL SERIES	CAPACITY
Aluminum	Pints
6, 8, 9 Cu. in. Vert. Crankshaft	1-1/4
6, 8, 9 Cu. in. Horiz. Crankshaft	1-1/4
10, 13 Cu. in. Vert. Crankshaft	1-3/4
10, 13 Cu. in. Horiz. Crankshaft	1-1/4
14, 17 Cu. in. Vert. Crankshaft	2-1/4
14, 17, 19 Cu. in. Horiz. Crankshaft	2-3/4
Cast Iron	
9, 14, 19, 20 Cu. in. Horiz. Crankshaft	3
23, 24, 30, 32 Cu. in. Horiz. Crankshaft	4

Crankcase capacity chart

internal bracket which will be distorted if pressure is applied to the fiber disc valve. Therefore, do not apply force when checking the valve with the wire gauge.

If the breather is removed for inspection or valve repair, a new gasket should be used when replacing the breather. Tighten the screws securely to prevent oil leakage.

Most breathers are now vented through the air cleaner, to prevent dirt from entering the crankcase. Check to be sure that the venting elbows or the tube are not damaged and that they are properly sealed.

OIL DIPPERS AND SLINGERS

Oil dippers reach into the oil reservoir in the base of the engine and splash oil onto the internal engine parts. The oil

dipper is installed on the connecting rod and has no pump or moving parts.

Oil slingers are driven by the cam gear. Old style slingers using a die cast bracket assembly have a steel bushing between the slinger and the bracket. Replace the bracket on which the oil slinger rides if it is worn to a diameter of 0.490 in. or less. Replace the steel bushing if it is worn. Newer style oil slingers have a stamped steel bracket.

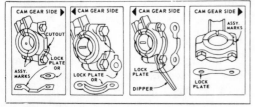

Installing the connecting rod in a horizontal crankshaft engine

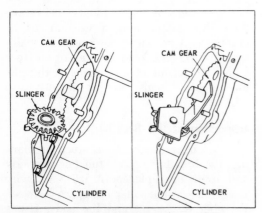

Installation of an oil slinger

Starters

REWIND STARTERS

Spring Removal and Installation

1. Cut the knot at the starter pulley to remove the rope.

2. With the rope removed, grasp the outer end of the starter spring with a pair of pliers and pull it out of the housing as far as possible.

3. Bend one of the bumper tangs up and lift out the starter pulley thereby disconnecting the spring.

4. If the old spring is to be reinstalled, before doing so, clean the spring in solvent

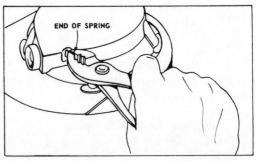

Removing the spring from a recoil starter

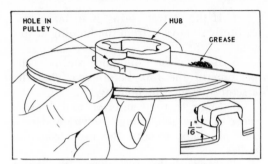
Installing the spring in a recoil starter

and wipe it clean by pulling it through a cloth. Straighten the spring to allow easier installation and to restore tension.

5. Insert either end of the spring into the blower housing slot and hook it into the pulley.

6. Replace the nylon bumpers if they are worn. Place a daub of grease on the pulley. Set the pulley into the housing and bend the bumper tang down.

7. To rewind the spring, place a ¾ in. square piece of stock into the center of the pulley hub. Grasping the stock with a wrench, wind the pulley until the spring is wound tight. Then back off the pulley one turn until the hole in the pulley for the rope knot and eyelet in the blower housing are in alignment. The spring should be securely locked in the small portion of the tapered hole.

8. Before installing the rope, inspect it for fraying. If it is frayed, replace it.

9. Insert the rope through the handle and tie a figure eight knot in the end. Insert the pin through the knot and pull it tightly into the handle. Always seal both ends of the knot. If you are re-using the old rope, burn the pulley end of the rope with a match. Wipe the end with a piece of waste cloth while it is still hot to prevent swelling and unravelling.

10. Inspect and clean the starter clutch assembly as necessary. Replace any worn parts. If necessary, the new type sealed clutch can be disassembled by using a screwdriver or wedge to pry open the retainer cover from the housing.

WINDUP STARTERS

There are two types of windup starters. The control KNOB release is used with the unsealed, four balled clutch. The control LEVER release can only be used with the sealed six balled clutch.

NOTE: *Before working on any windup starter, make sure the starter spring is not wound. This can be determined by attempting to turn the starter crank clockwise. If it is wound tight, release the tension by placing the control knob or lever to the "Start" position. If the starter spring does not release, place the control knob or lever at the "Crank" position. To prevent injury, hold the crank handle with one hand while removing the phillips head screw and handle assembly from the starter housing. This will release the spring.*

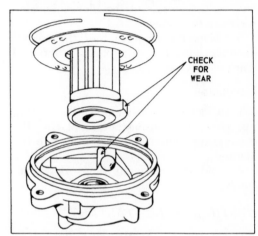
Disassembling the starter clutch in a windup starter

Disassembly

If you suspect that the starter has a broken spring, check before removing it from the engine. Place the control knob or lever in the "Start" position. Turn the cranking handle ten turns clockwise. If the engine does not turn over, either the spring is broken or the starter clutch balls are not engaging. While turning the cranking handle, watch the starter clutch ratchet. If it does not move, the starter spring is probably broken.

1. Remove the blower housing.

2. Remove the screw which holds the cranking handle to the housing.

3. Bend the tangs which hold the starter spring and housing assembly upward and lift the retainer plate, spring, and housing assembly out of the blower housing.

CAUTION: *Do not attempt to remove the starter spring from its housing.*

4. Inspect the spring and housing as-

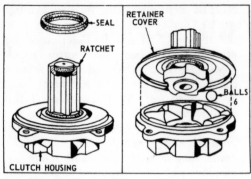

A sealed clutch assembly

sembly for damage. Inspect the ratchet gear on the outside of the blower housing for wear or damage.

5. Check the movement of the control knob or lever for ease of operation and damage or wear.

Installation

1. Be sure to reassemble the spring washer in the housing before placing the cup, spring, and release assembly into the housing.

2. Bend the retaining tangs down securely.

VERTICAL PULL STARTERS

Rope and/or Spring Removal and Installation

NOTE: *Before servicing the starter, remove all tension from the rope.*

1. After removing the starter assembly from the engine, use a screwdriver to lift the rope up approximately 1 ft.

2. Wind the rope and pulley counterclockwise two or three turns. This will completely release tension from the starter spring.

CAUTION: *Note the warning on the plastic cover, then use a screwdriver to remove the cover. Do not pull the rope with the pulley cover removed unless the spring is detached from the spring anchor.*

3. Remove the anchor bolt and anchor. Inspect the starter spring for kinks or damaged ends.

4. If the starter spring is to be reused, carefully remove it from the housing at this time.

5. Remove the rope guide and note the position of the link before removing the assembly from the housing.

6. Make a rope inserter tool from a piece of wire and, using the wire and/or a pair of pliers, remove the rope from the pulley.

7. Remove the rope from the hand grip. Inspect the rope to see if it is frayed or worn.

8. It is not necessary to remove the gear retainer unless the pulley or gear is damaged and will be replaced.

9. Clean all dirty or oily parts and check them for proper friction. The link should move the gear in both extremes of its travel. If it does not, replace the link assembly.

10. Install the spring by hooking the end in the pulley retainer slot and winding until the spring is coiled in the housing.

11. Thread the rope through the grip and into the insert. Tie a small, tight knot in the end. Heat seal the knot to prevent loosening, Pull the knot into the insert pocket and snap the insert into the grip.

12. Insert the rope through the housing and into the pulley using the rope inserter (wire). Tie a small knot, heat seal it, and pull it tight into the recess in the rope pulley. The rope must not interfere with the gear motion.

13. Install the pulley assembly in the housing with the link in the pocket of the casting. Install the rope guide.

14. Rotate the pulley in a counterclockwise direction until the rope is fully retrieved.

15. Hook the free end of the spring to the spring anchor and install the screw. Lubricate the spring with a small quantity of engine oil.

16. Snap the cover in place. Wind the starter spring by pulling the rope out approximately 1 ft. Wind the rope and pulley two or three turns clockwise to achieve the proper rope tension.

3 · Tecumseh-Lauson

4 Stroke Engines

ENGINE IDENTIFICATION

Lauson 4 cycle engines are identified by a model number stamped on a name-plate. The nameplate is located on the crankcase of vertical shaft models and on the blower housing of horizontal shaft models.

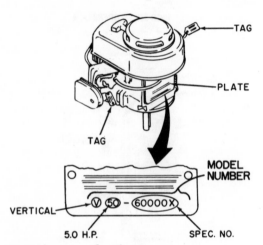

Vertical engine identification

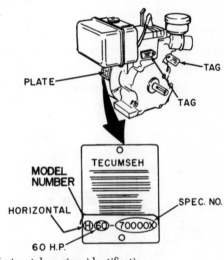

Horizontal engine identification

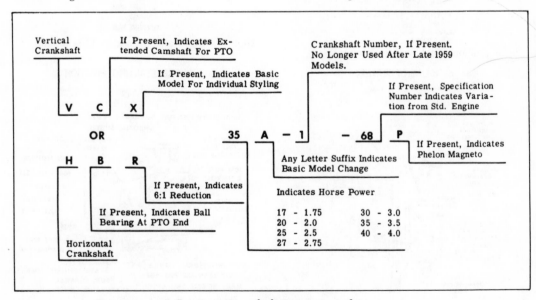

Interpretation of the engine number

AIR CLEANERS

Service all of the oil/foam polyurethane and oil bath air cleaner elements in the same manner as the Briggs and Stratton components. See Chapter Two.

The Tecumseh treated paper element type air cleaner consists of a pleated paper element encased in a metal housing and must be replaced as a unit. A flexible tubing and hose clamps connect the remotely mounted air filter to the carburetor.

Clean the element by lightly tapping it. Do not distort the case. When excessive carburetor adjustment or loss of power results, inspect the air filter to see if it is clogged. Replacing a severely restricted air filter should show an immediate performance improvement.

GENERAL CARBURETOR SERVICE

Four-cycle Tecumseh engines use float or diaphragm type carburetors.

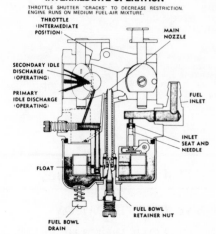

INTERMEDIATE OPERATION

THROTTLE SHUTTER "CRACKS" TO DECREASE RESTRICTION. ENGINE RUNS ON MEDIUM FUEL-AIR MIXTURE.

Intermediate operation—float feed carburetor

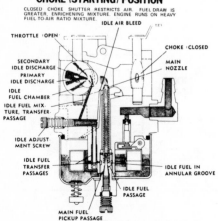

CHOKE (STARTING) POSITION

CLOSED CHOKE SHUTTER RESTRICTS AIR. FUEL DRAW IS GREATER, ENRICHENING MIXTURE. ENGINE RUNS ON HEAVY FUEL-TO-AIR RATIO MIXTURE.

Choke position—float feed carburetor

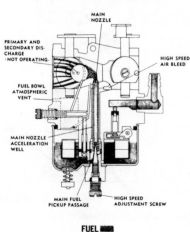

HIGH SPEED OPERATION

VENTURI REPLACES THROTTLE SHUTTER AS THE RESTRICTING DEVICE. ENGINE RUNS ON FULL POWER FUEL-AIR MIXTURE.

FUEL
AIR
MIXTURE

High speed operation—float feed carburetor

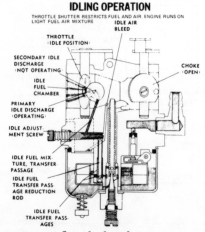

IDLING OPERATION

THROTTLE SHUTTER RESTRICTS FUEL AND AIR. ENGINE RUNS ON LIGHT FUEL-AIR MIXTURE.

Idle operation—float feed carburetor

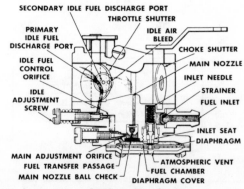

CHOKE (STARTING) POSITION

CLOSED CHOKE SHUTTER RESTRICTS AIR. FUEL DRAW IS GREATER, ENRICHENING MIXTURE. ENGINE RUNS ON HEAVY FUEL-TO-AIR RATIO MIXTURE.

Choke position—diaphragm carburetor

IDLING OPERATION

THROTTLE SHUTTER RESTRICTS FUEL AND AIR. ENGINE RUNS ON LIGHT POWER FUEL-AIR MIXTURE.

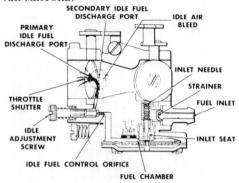

Idling position—diaphragm carburetor

INTERMEDIATE OPERATION

THROTTLE SHUTTER "CRACKS" TO DECREASE RESTRICTION. ENGINE RUNS ON MEDIUM POWER FUEL-AIR MIXTURE.

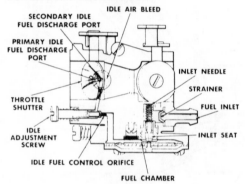

Intermediate operation—diaphragm carburetor

HIGH SPEED OPERATION

VENTURI REPLACES THROTTLE SHUTTER AS THE RESTRICTING DEVICE. ENGINE RUNS ON FULL POWER FUEL AIR MIXTURE.

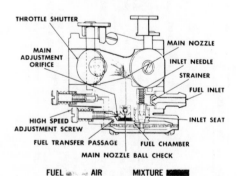

High speed operation—diaphragm carburetor

Fuel Mixture Adjustments

Check the adjustment screw tip for damage. If a ridge ring on the seat area can be felt with your thumb nail, the needle must be replaced.

Three screws must be initially adjusted before attempting to operate a newly overhauled carburetor:

Idle Mixture and Main Mixture Screw

Turn the screw completely in until finger tight and then back it out one full turn.

Idle Speed Screw

Back out the screw. Then turn it in until the screw just touches the throttle lever and continue one turn more.

NOTE: *Some carburetors have a fixed main mixture and/or idle mixture jets. The absence of the adjustment screw and receptacle indicate a fixed jet, and therefore no adjustment is necessary.*

Final Carburetor Adjustments

1. Allow the engine to warm up to normal operating temperature.

2. With the engine running at maximum recommended rpm, loosen the main metering screw until the engine starts to "lope" or roll, then tighten the screw until the engine starts to cut out.

3. Note the number of turns from one extreme to the other. Loosen the screw to a point midway between the extremes.

4. Test the engine by running it under normal load conditions. The engine should respond to load pick up immediately. An engine that "dies" is too lean. A lean engine may knock, indicating that a float adjustment is necessary.

An engine which ran roughly before picking up the load is adjusted too rich.

NOTE: *If the adjustment seems too touchy, check the float for proper setting and for sticking.*

FLOAT FEED CARBURETOR

Identification

When identifying the carburetor, use the engine model number and the number that is stamped on the carburetor body.

Rebuilding

Remove the carburetor from the engine. It is easier to remove the intake manifold and carburetor assembly from the engine, disconnect the governor linkage, fuel line, and grounding wire and then disassemble the carburetor from the intake manifold on a work bench. Be sure to note the posi-

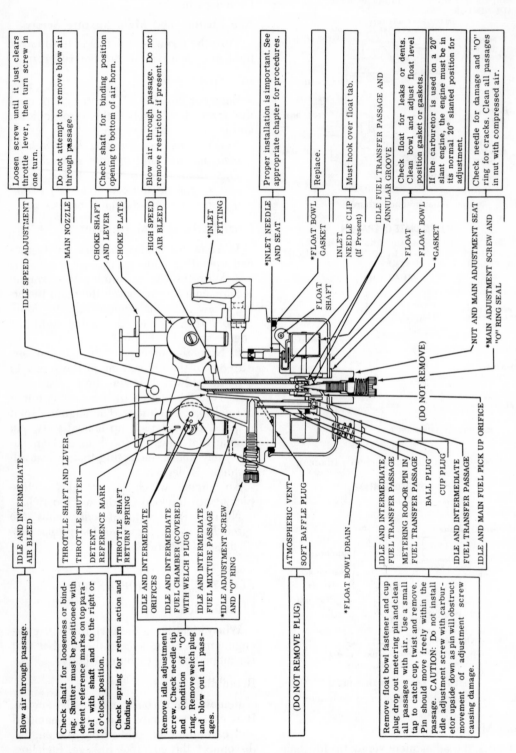

Service hints for float feed carburetors

Blow air through passage.

IDLE AND INTERMEDIATE
AIR BLEED

IDLE SPEED ADJUSTMENT

Loosen screw until it just clears throttle lever, then turn screw in one turn.

MAIN NOZZLE

Do not attempt to remove blow air through passage.

CHOKE SHAFT
AND LEVER

Check shaft for binding position opening to bottom of air horn.

CHOKE PLATE

HIGH SPEED
AIR BLEED

Blow air through passage. Do not remove restrictor if present.

*INLET
FITTING

*INLET NEEDLE
AND SEAT

Proper installation is important. See appropriate chapter for procedures.

*FLOAT BOWL
GASKET

Replace.

INLET
NEEDLE CLIP
(If Present)

Must hook over float tab.

IDLE FUEL TRANSFER PASSAGE AND
ANNULAR GROOVE

FLOAT

Check float for leaks or dents. Clean bowl and adjust float level position gasket or gaskets.

*FLOAT BOWL

If the carburetor is used on a 20° slant engine, the engine must be in its normal 20° slanted position for adjustment.

*GASKET

NUT AND MAIN ADJUSTMENT SEAT

*MAIN ADJUSTMENT SCREW AND
"O" RING SEAL

Check needle for damage and "O" ring for cracks. Clean all passages in nut with compressed air.

Check shaft for looseness or binding. Shutter must be positioned with detent reference marks on top parallel with shaft and to the right or 3 o'clock position.

THROTTLE SHAFT AND LEVER

THROTTLE SHUTTER

DETENT
REFERENCE MARK

Check spring for return action and binding.

THROTTLE SHAFT
RETURN SPRING

IDLE AND INTERMEDIATE
ORIFICES

Remove idle adjustment screw. Check needle tip and condition of "O" ring. Remove welch plug and blow out all passages.

IDLE AND INTERMEDIATE
FUEL CHAMBER (COVERED
WITH WELCH PLUG)

IDLE AND INTERMEDIATE
FUEL MIXTURE PASSAGE

*IDLE ADJUSTMENT SCREW
AND "O" RING

FLOAT
SHAFT

(DO NOT REMOVE)

ATMOSPHERIC VENT

SOFT BAFFLE PLUG

(DO NOT REMOVE PLUG)

*FLOAT BOWL DRAIN

IDLE AND INTERMEDIATE
FUEL TRANSFER PASSAGE

METERING ROD-OR PIN IN
FUEL TRANSFER PASSAGE

BALL PLUG

CUP PLUG

IDLE AND INTERMEDIATE
FUEL TRANSFER PASSAGE

IDLE AND MAIN FUEL PICK UP ORIFICE

Remove float bowl fastener and cup plug drop out metering pin and clean all passages with air. Use a small tap to catch cup, twist and remove. Pin should move freely within the passage. CAUTION: Do not install idle adjustment screw with carburetor upside down as pin will obstruct movement of adjustment screw causing damage.

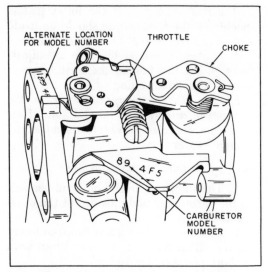

Float feed carburetor identification number

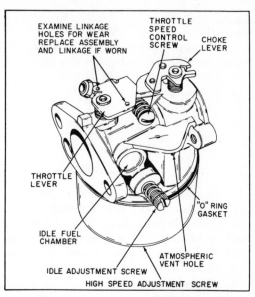

External view of a Tecumseh float type carburetor

tions of the governor and the throttle linkage to facilitate reassembly.

THROTTLE

1. Examine the throttle lever and plate prior to disassembly. Replace any worn parts.

2. Remove the screw in the center of the throttle plate and pull out the throttle shaft lever assembly.

3. When reassembling, it is important that the lines on the throttle plate are facing out when in the closed position. Position the throttle plates with the two lines at 12 and 3 o'clock. The throttle shaft must be held in tight to the bottom bearing to prevent the throttle plate from riding on the throttle bore of the body which would cause excessive throttle plate wear and governor hunting.

CHOKE

Examine the choke lever and shaft at the bearing points and holes into which the linkage is fastened and replace any worn parts. The choke plate is inserted into the air horn of the carburetor in such a way that the flat surface of the choke is toward the fuel bowl.

IDLE ADJUSTING SCREW

Remove the idle screw from the carburetor body and examine the point for damage to the seating surface on the taper. If damaged, replace the idle adjusting needle. Tension is maintained on the screw with a

coil spring and sealed with an O-ring. Examine and replace the O-ring if it is worn or damaged.

HIGH SPEED ADJUSTING JET

Remove the screw and examine the taper. If the taper is damaged at the area where it seats, replace the screw and fuel bowl retainer nut as an assembly.

The fuel bowl retainer nut contains the seat for the screw. Examine the sealing O-ring on the high speed adjusting screw. Replace the O-ring if it indicates wear or cuts. During the reassembly of the high speed adjusting screw, position the coil spring on the adjusting screw, followed by the small brass washer and the O-ring seal.

FUEL BOWL

To remove the fuel bowl, remove the retaining nut and fiber washer. Replace the nut if it is cracked or worn.

The retaining nut contains the transfer passage through which fuel is delivered to the high speed and idle fuel system of the carburetor. It is the large hole next to the hex nut end of the fitting. If a problem occurs with the idle system of the carburetor, examine the small fuel passage in the annular groove in the retaining nut. This passage must be clean for the proper transfer of fuel into the idle metering system.

The fuel bowl should be examined for rust and dirt. Thoroughly clean it before installing it. If it is impossible to properly clean the fuel bowl, replace it.

Check the drain valve for leakage. Replace the rubber gasket on the inside of the drain valve if it leaks.

Examine the large O-ring that seals the fuel bowl to the carburetor body. If it is worn or cracked, replace it with a new one, making sure the same type is used (square or round).

FLOAT

1. Remove the float from the carburetor body by pulling out the float axle with a pair of needle nose pliers. The inlet needle will be lifted off the seat because it is attached to the float with an anchoring clip.

2. Examine the float for damage and holes. Check the float hinge for wear and replace it if worn.

3. The float level is checked by positioning a #4 (0.209 in.) twist drill across the rim between the center leg and the unmachined surface of the index pad, parallel

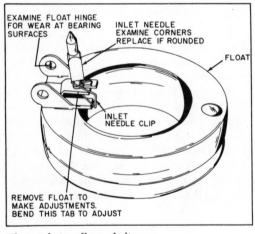

Float, inlet needle, and clip

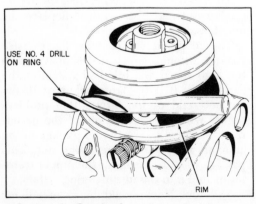

Adjusting the float level

to the float axle pin. If the index pad is machined, the float setting should be made with a #9 (0.180–0.200 in.) twist drill.

4. Remove the float to make an adjustment. Bend the tab on the float hinge to correct the float setting.

NOTE: *Direct compressed air in the opposite direction of normal flow of air or fuel (reverse taper) to dislodge foreign matter.*

INLET NEEDLE AND SEAT

1. The inlet needle sits on a rubber seat in the carburetor body instead of the usual metal fitting.

2. Remove it, place a few drops of heavy engine oil on the seat, and pry it out with a short piece of hooked wire.

3. The grooved side of the seat is inserted first. Lubricate the cavity with oil and use a flat faced punch to press the inlet seat into place.

4. Examine the inlet needle for wear and rounding off of the corners. If this condition does exist, replace the inlet needle.

FUEL INLET FITTING

1. The inlet fitting is removed by twisting and pulling at the same time.

2. Use sealer when reinstalling the fitting. Insert the tip of the fitting into the carburetor body. Press the fitting in until the shoulder contacts the carburetor. Only use inlet fittings without screens.

CARBURETOR BODY

1. Check the carburetor body for wear and damage.

2. If excessive dirt has accumulated in the atmospheric vent cavity, try cleaning it with carburetor solvent or compressed air. Remove the welch plug only as a last resort.

NOTE: *The carburetor body contains a pressed-in main nozzle tube at a specific depth and position within the venturi. Do not attempt to remove the main nozzle. Any change in nozzle positioning will adversely affect the metering quality and will require carburetor replacement.*

3. Clean the accelerating well around the main nozzle with compressed air and carburetor cleaning solvents.

4. The carburetor body contains two cup plugs, neither of which should be removed. A cup plug located near the inlet seat

cavity, high up on the carburetor body, seals off the idle bleed. This is a straight passage drilled into the carburetor throat. Do not remove this plug. Another cup plug is located in the base where the fuel bowl nut seals the idle fuel passage. Do not remove this plug or the metering rod.

5. A small ball plug located on the side of the idle fuel passage seals this passage. Do not remove this ball plug.

6. The welch plug on the side of the carburetor body, just above the idle adjusting screw, seals the idle fuel chamber. This plug can be removed for cleaning of the idle fuel mixture passage and the primary and secondary idle fuel discharge ports. Do not use any tools that might change the size of the discharge ports, such as wire or pins.

RESILIENT TIP NEEDLE

Replace the inlet needle. Do not attempt to remove or replace the seat in the carburetor body.

VITON SEAT

Using a 10–24 or 10–32 tap, turn the tap into the brass seat fitting until it grasps the seat firmly. Clamp the tap shank into a vise and tap the carburetor body with a soft hammer until the seat slides out of the body.

To replace the viton seat, position the replacement over the receptical with the soft rubber like seat toward the body. Use a flat punch and a small hammer to drive the seat into the body until it bottoms on the shoulder.

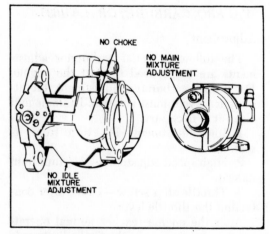

Auto-Magic carburetor

TECUMSEH AUTOMATIC NON-ADJUSTABLE FLOAT FEED CARBURETOR

This carburetor has neither a choke plate nor idle and main mixture adjusting screws. There is no running adjustment. The float adjustment is the standard Tecumseh float setting of 0.210 in. (#4 drill).

Cleaning

Remove all non-metallic parts and clean them using a procedure similar to that for the other carburetors. Never use wires through any of the drilled holes. Do not remove the baffling welch plug unless it is certain there is a blockage under the plug. There are no blind passageways in this carburetor.

WALBRO AND TILLOTSON FLOAT FEED CARBURETORS

Procedures are similar to those for the Tecumseh float carburetor with the exceptions noted below.

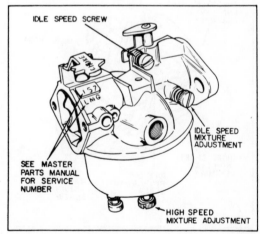

Walbro float feed carburetor

Rebuilding

MAIN NOZZLE

The main nozzle in Walbro carburetors is cross drilled after it is installed in the carburetor. Once removed, it cannot be reinstalled, since it is impossible to properly realign the cross drilled holes. Grooved service replacement main nozzles are available which allow alignment of these holes.

Float Shaft Spring

Carefully position the float shaft spring on models so equipped. The spring dampens float action when properly assembled. Use needle-nosed pliers to hook the end of the spring over the float hinge and then insert the pin as far as possible before lifting the spring from the hinge into position. Leaving the spring out or improper installation will cause unbalanced float action and result in a touchy adjustment.

Float Adjustment

1. To check float adjustment, invert the assembled float carburetor body. Check the clearance between the body and the float, opposite the hinge. Clearance should be $\frac{1}{8}$ in. $\pm$ $\frac{1}{64}$ in.

2. To adjust the float level, remove the float shaft and float. Bend the lip of the float tang to correct the measurement.

3. Assemble the parts and recheck the adjustment.

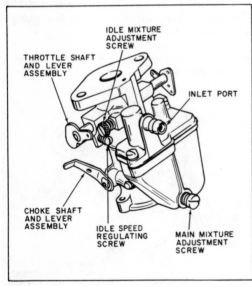

IDLE MIXTURE ADJUSTMENT SCREW

THROTTLE SHAFT AND LEVER ASSEMBLY

INLET PORT

CHOKE SHAFT AND LEVER ASSEMBLY

IDLE SPEED REGULATING SCREW

MAIN MIXTURE ADJUSTMENT SCREW

Tillotson Model E float carburetor

Tillotson E Float Feed Type Carburetor

The following adjustments are different for this carburetor.

Running Adjustment

1. Start the engine and allow it to warm up to operating temperatures. Make sure the choke is fully opened after the engine is warmed up.

2. Run the engine at a constant speed while slowly turning the main adjustment screw in until the engine begins to lose speed; then slowly back it out about $\frac{1}{8}$–$\frac{1}{4}$ of a turn until maximum speed and power is obtained (4000 rpm). This is the correct power adjustment.

3. Close the throttle and cause the engine to idle slightly faster than normal by turning the idle speed regulating screw in. Then turn the idle adjustment speed screw in until the engine begins to lose speed; then turn it back $\frac{1}{4}$ to $\frac{1}{2}$ of a turn until the engine idles smoothly. Adjust the idle speed regulating screw until the desired idling speed is acquired.

4. Alternately open and close the throttle a few times for an acceleration test. If stalling occurs at idle speeds, repeat the adjustment procedures to get the proper idle speed.

Float Level Adjustment

1. Remove the carburetor float bowl cover and float mechanism assembly.

2. Remove the float bowl cover gasket and, with the complete assembly in an upside down position and the float lever tang resting on the seated inlet needle, a measurement of $1\frac{5}{64}$ in. should be maintained from the free end flat rim, or edge of the cover, to the toe of the float. Measurement can be checked with a standard straight rule or depth gauge.

3. If it is necessary to raise or lower the float lever setting, remove the float lever pin and the float, then carefully bend the float lever tang up or down as required to obtain the correct measurement.

WALBRO CARBURETOR #631635

Adjustment

The following initial carburetor adjustments are to be used to start the engine. For proper carburetion adjustment, the atmospheric vent must be open. Examine and clean it if necessary.

1. Idle adjustment—$1\frac{1}{4}$ turn from its seat.

2. High speed adjustment—$1\frac{1}{2}$ turn from its seat.

3. Throttle stop screw—1 turn after contacting the throttle lever.

After the engine reaches normal operating temperature, make the final adjustments

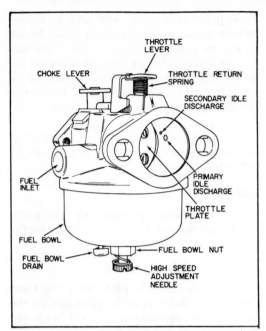

Walbro Carburetor #631635—engine side

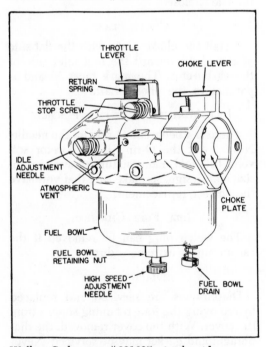

Walbro Carburetor #631635—intake side

for best idle and high speed within the following ranges. Recommended speeds: Idle: 1800–2300 rpm. High speed: 3450–3750 rpm.

Rebuilding Notes

1. The throttle plate is installed with the lettering (if present) facing outward when closed. The throttle plate is installed on the throttle lever with the lever in the closed position if there is binding after the plate is in position, loosen the throttle plate and reposition it.

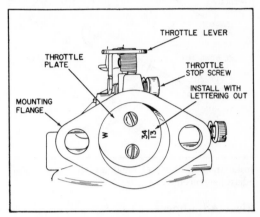

Installation of the throttle plate

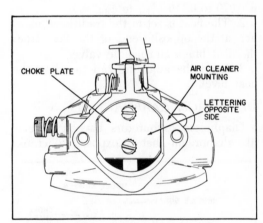

Installation of the choke plate

2. Before removing the fuel bowl nut, remove the high speed adjusting needle. Use a $7/16$ in. box wrench or socket to remove the fuel bowl nut. When replacing the fuel bowl nut, be sure to position the fiber gasket under the nut and tighten it securely.

3. Examine the high speed needle tip and, if it appears to be worn, replace it. When the high speed jet is replaced, the main nozzle, which includes the jet seat, should also be replaced. The original main nozzle cannot be used. There are special replacement nozzles available.

4. The inlet needle valve is replaceable if it appears to be worn. The inlet valve seat is also replaceable and should be replaced if the needle valve is replaced.

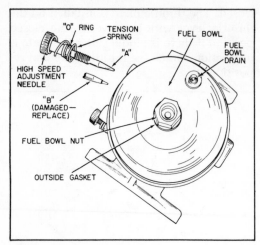

Installation of the fuel bowl and the high speed adjustment needle

Float Adjustment

1. The float setting for this carburetor is 0.070 to 0.110 ($\frac{5}{64}$ to $\frac{7}{64}$ in.).

2. The float is set in the traditional manner, at the opposite end of the float from the float hinge and needle valve.

3. Bend the adjusting tab to adjust the float level.

DIAPHRAGM CARBURETORS

Diaphragm carburetors have a rubber-like diaphragm that is exposed to crank-

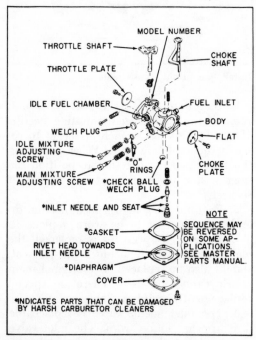

Exploded view of a Tecumseh diaphragm carburetor

case pressure on one side and to atmospheric pressure on the other side. As the crankcase pressure decreases, the diaphragm moves against the inlet needle allowing the inlet needle to move from its seat which permits fuel to flow through the inlet valve to maintain the correct fuel level in the fuel chamber.

An advantage of this type of system over the float system, is that the engine can be operated in any position.

Rebuilding

Use carburetor cleaner only on metal parts, except for the main nozzle in the main body.

THROTTLE PLATE

Install the throttle plate with the short line that is stamped in the plate toward the top of the carburetor, parallel with the throttle shaft, and facing out when the throttle is closed.

CHOKE PLATE

Install the choke plate with the flat side of the choke toward the fuel inlet side of the carburetor. The mark faces in and is parallel to the choke shaft.

IDLE MIXTURE ADJUSTMENT SCREW

There is a neoprene O-ring on the needle. Never soak the O-ring in carburetor solvent. Idle and main mixture screws vary in size and design, so make sure that you have the correct replacement.

IDLE FUEL CHAMBER

The welch plug can be removed if the carburetor is extremely dirty.

DIAPHRAGMS

Diaphragms are serviced and replaced by removing the four retaining screws from the cover. With the cover removed, the diaphragm and gasket may be serviced. Never soak the diaphragm in carburetor solvent. Replace the diaphragm if it is cracked or torn. Be sure there are no wrinkles in the diaphragm when it is replaced. The diaphragm rivet head is always placed facing the inlet needle valve.

INLET NEEDLE AND SEAT

The inlet seat is removed by using either a slotted screwdriver (early type) or a $\frac{9}{32}$

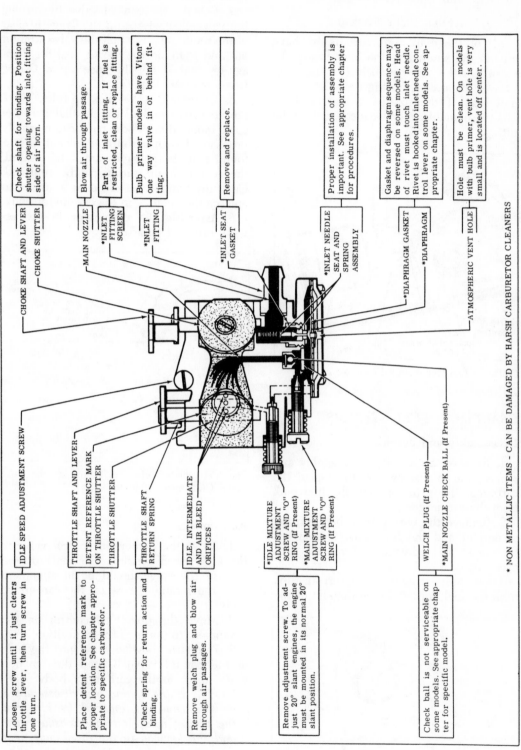

CHOKE SHAFT AND LEVER — Check shaft for binding. Position shutter opening towards inlet fitting side of air horn.

CHOKE SHUTTER

MAIN NOZZLE — Blow air through passage.

*INLET FITTING SCREEN — Part of inlet fitting. If fuel is restricted, clean or replace fitting.

INLET FITTING — Bulb primer models have Viton one way valve in or behind fitting.

*INLET SEAT GASKET — Remove and replace.

*INLET NEEDLE SEAT AND SPRING ASSEMBLY — Proper installation of assembly is important. See appropriate chapter for procedures.

*DIAPHRAGM GASKET — Gasket and diaphragm sequence may be reversed on some models. Head of rivet must touch inlet needle. Rivet is hooked into inlet needle control lever on some models. See appropriate chapter.

*DIAPHRAGM

ATMOSPHERIC VENT HOLE — Hole must be clean. On models with bulb primer, vent hole is very small and is located off center.

IDLE SPEED ADJUSTMENT SCREW — Loosen screw until it just clears throttle lever, then turn screw in one turn.

THROTTLE SHAFT AND LEVER — Place detent reference mark to proper location. See chapter appropriate to specific carburetor.

DETENT REFERENCE MARK ON THROTTLE SHUTTER

THROTTLE SHUTTER

THROTTLE SHAFT RETURN SPRING — Check spring for return action and binding.

IDLE, INTERMEDIATE AND AIR BLEED ORIFICES — Remove welch plug and blow air through air passages.

*IDLE MIXTURE ADJUSTMENT SCREW AND "O" RING (If Present)

*MAIN MIXTURE ADJUSTMENT SCREW AND "O" RING (If Present) — Remove adjustment screw. To adjust 20° slant engines, the engine must be mounted in its normal 20° slant position.

WELCH PLUG (If Present)

*MAIN NOZZLE CHECK BALL (If Present) — Check ball is not serviceable on some models. See appropriate chapter for specific model.

* NON METALLIC ITEMS - CAN BE DAMAGED BY HARSH CARBURETOR CLEANERS

Service hints for the diaphragm carburetors

in. socket. The inlet needle is spring loaded, so be careful when removing it.

FUEL INLET FITTING

All of the diaphragm carburetors have an integral strainer in the inlet fitting. To clean it, either reverse flush it or use compressed air after removing the inlet needle and seat. If the strainer is lacquered or otherwise unable to be cleaned, replace the fitting.

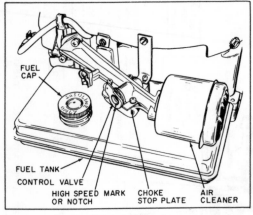

Craftsman fuel tank mounted carburetor

CRAFTSMAN FUEL SYSTEMS

Disassembly and Service

1. Remove the air cleaner assembly and remove the four screws on the top of the carburetor body to separate the fuel tank from the carburetor.

2. Remove the O-ring from between the carburetor and the fuel tank. Examine it for cracks and damage and replace it if necessary.

3. Carefully remove the reservoir tube from the fuel tank. Observe the end of the

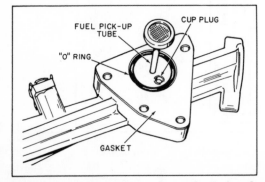

Positioning of the reservoir tube in the fuel tank

tube that rested on the bottom of the fuel tank. It should be slotted.

4. Remove the control valve by turning the valve clockwise until the flange is clear of the retaining boss. Pull the valve straight out and examine the O-ring seal for damage or wear. If possible, use a new O-ring when reassembling.

5. Examine the fuel pick up tube. There are no valves or ball checks that may become inoperative. These parts can normally be cleaned with carburetor solvent. If it is found that the passage cannot be cleared, the fuel pick up tubes can be replaced. Carefully remove the old ones so as not to enlarge the opening in the carburetor body.

6. Assemble the carburetor in the reverse order of disassembly.

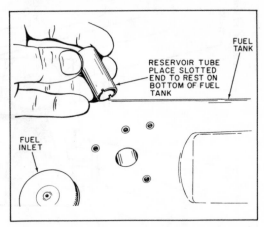

The fuel pick-up tube and O-ring on the early model Craftsman carburetor

Adjustments

1. Move the carburetor control valve to the high speed position. The mark on the face of the valve should be in alignment with the retaining boss on the carburetor body.

2. Move the operator's control on the equipment to the high speed position.

3. Insert the bowden wire into the hole of the control valve. Clamp the bowden wire sheath to the carburetor body.

NOTE: *If the engine was disassembled and the camshaft removed, be sure that the timing marks on the camshaft gear and the keyway in the crankshaft gear are aligned when reinserting the camshaft. Then lift the camshaft enough to advance the camshaft gear timing mark to the right (clockwise) ONE tooth, as*

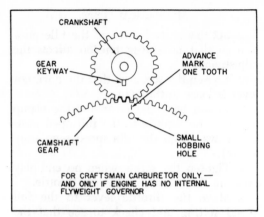

Timing an engine with the Craftsman fuel tank mounted carburetor

viewed from the power take-off end of the crankshaft.

CRAFTSMAN FLOAT TYPE CARBURETORS

Craftsman float type carburetors are serviced in the same manner as the other Tecumseh float type carburetors. The throttle control valve has three positions: stop, run, and start.

When the control valve is removed, replace the O-ring. If the engine runs sluggishly, consider the possibility of a leaky O-ring.

Timing for engines that use these car-

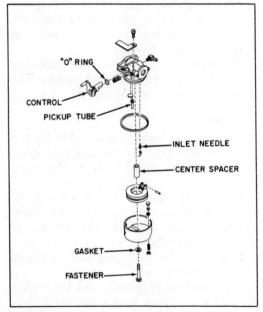

Exploded view of the Craftsman float type carburetor

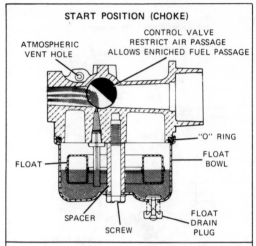

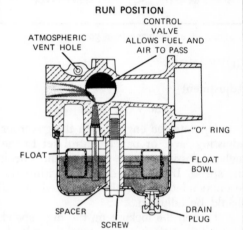

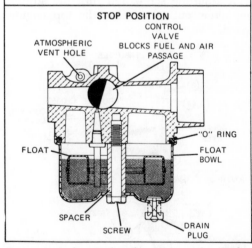

Start, run and stop throttle positions of a Craftsman float type carburetor

buretors is the same as the other Tecumseh engines, that is, align the timing marks on the camshaft gear and the crankshaft. Never advance the camshaft timing mark.

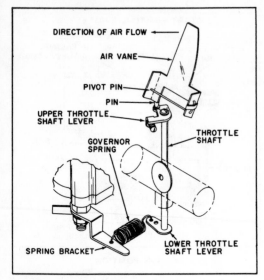

Diagram of an air vane type governor

GOVERNORS

Adjustment

AIR VANE TYPE

1. Operate the engine with the governor adjusting lever or panel control set to the highest possible speed position and check the speed. If the speed is not within the recommended limits, the governed speed should be adjusted.

2. Loosen the locknut on the high speed limit adjusting screw and turn the adjusting screw out to increase the top engine speed.

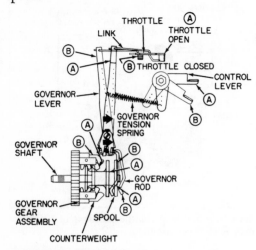

Ⓐ GOVERNED POSITION

Ⓑ NON-GOVERNED POSITION

➡ DIRECTION OF ADJUSTMENT – THIS SCHEMATIC

Diagram of a mechanical type governor

MECHANICAL TYPE

1. Set the control lever to the idle position so that no spring tension affects the adjustment.

2. Loosen the screw so that the governor lever is loose in the clamp.

3. Rotate both the lever and the clamp to move the throttle to the full open position (away from the idle speed regulating screw).

4. Tighten the screw when no end-play exists in the direction of open throttle.

5. Move the throttle lever to the full speed setting and check to see that the control linkage opens the throttle.

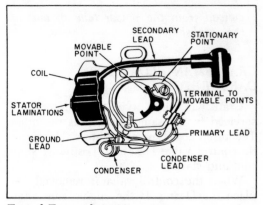

Typical Tecumseh ignition system

IGNITION SYSTEM

All Tecumseh engines have a magneto ignition. It consists of a stator assembly mounted to the engine and a magnet cast into the rotating flywheel.

The contact and condenser are under the flywheel, near the coil and stator. The points are crankshaft operated.

Checking

1. Grasp the high tension cable by the insulation and hold the terminal end about $\frac{1}{8}$ in. from the metal body of the spark plug.

2. Crank the engine, and if a hot spark jumps the gap between the center electrode and the ground electrode, the magneto is operating correctly.

3. Remove the spark plug and reconnect the high tension lead.

4. Ground the plug and crank the engine over. If a hot spark jumps the spark gap, the ignition system is operating satisfactorily.

Magneto Adjustment

1. Disconnect the fuel line from the carburetor.

2. Remove the mounting screws, fuel tank, and shroud to provide access to the flywheel.

3. Remove the flywheel with either a puller (over 3.5 hp) or by using a screwdriver to pry underneath the flywheel while tapping the top lightly with a hammer.

4. Remove the dust cover and gasket from the magneto and crank the engine over until the breaker points of the magneto are fully opened.

5. Check the condition of the points and replace them if they are burned or pitted.

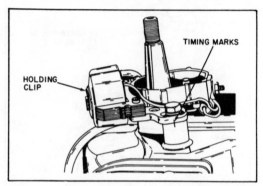

Typical timing marks for the Tecumseh ignition system

6. Check the point gap with a feeler gauge. Adjust them, if necessary, as per the directions on the dust cover. Refer to the specifications chart at the end of this chapter for point gap.

SOLID STATE IGNITION SYSTEM

The solid state ignition system has no friction (thus, wearing) components. The only moving part of the system is the flywheel with the charging magnets. The solid state ignition system functions in the following manner. As the engine flywheel magnet passes the input coil, a low voltage AC current is induced in that coil. The current passes through a rectifier which converts it to DC. It then travels to the capacitor where it is stored. The flywheel rotates about 180° and, as it passes the trigger coil, it induces a very small electric charge into that coil. The charge passes through the resistor and turns on the silicon controlled rectifier (solid state switch).

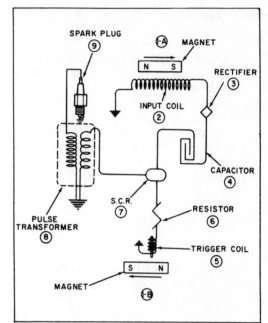

Wiring diagram for the solid state ignition system

With the silicon controlled rectifier closed, the low voltage stored in the capacitor travels to the pulse transformer. Here the voltage is stepped up instantaneously and it is discharged across the electrodes of the spark plug, firing just before top dead center. Some units are equipped with an advance and retard feature. This is accomplished through the use of a second trigger coil and resistor set to turn on the silicon controlled rectifier at a lower rpm to fire the spark plug at top dead center.

Checking

The only on-engine check which can be made to determine whether the ignition system is working, is to separate the high tension lead from the spark plug and check for spark. If there is a spark, then the unit is alright and the spark plug should be replaced. No spark indicates that some other part needs replacing.

Check the individual components as follows:

High Tension Lead—Inspect for cracks or indications of arcing. Replace the transformer if the condition of the lead is questionable.

Low Tension Leads—Check all leads for shorts. Check the ignition cut-off lead to see that the unit is not grounded. Repair the leads, if possible, or replace them.

Pulse Transformer—Replace and test for spark.

Magneto—Replace and test for spark. Time the magneto by turning it counter clockwise as far as it will go and then tighten the retaining screws.

Flywheel—Check the magnets for strength. With the flywheel off the engine, it should attract a screwdriver that is held 1 in. from the magnetic surface on the inside of the flywheel. Be sure that the key locks the flywheel to the crankshaft.

ENGINE MECHANICAL

Timing Gears

Correctly matched camshaft gear and crankshaft gear timing marks are necessary for the engine to perform properly.

On all camshafts the timing mark is located in line with the center of the hobbing hole (small hole in the face of the gear). If no line is visible, use the center of the hobbing hole to align with the crankshaft gear marked tooth.

On crankshafts where the gear is held on by a key, the timing mark is the tooth in line with the keyway.

On crankshafts where the gear is pressed onto the crankshaft, a tooth is bevelled to serve as the timing mark.

On engines with a ball bearing on the power take-off end of the crankshaft, look for a bevelled tooth which serves as the crankshaft gear timing mark.

If the engine uses a Craftsman type carburetor, the camshaft timing mark must be advanced clockwise one tooth ahead of the matching timing mark on the crankshaft, the exception being the Craftsman variable governed fuel systems.

NOTE: *If one of the timing gears, either the crankshaft gear or the camshaft gear, is damaged and has to be replaced, both gears should be replaced.*

Pistons

When removing the pistons, clean the carbon from the upper cylinder bore and head. The piston and pin must be replaced in matched pairs.

A ridge reamer must be used to remove the ridge at the top of the cylinder bore on some engines.

Clean the carbon from the piston ring

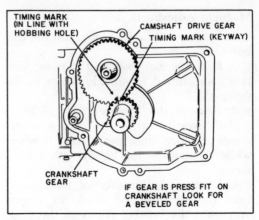

Timing marks

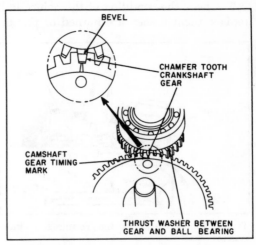

Timing marks

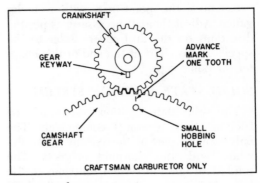

Timing marks

groove. A broken ring can be used for this operation.

Some engines have oversize pistons which can be identified by the oversize engraved on the piston top.

There is a definite piston-to-connecting rod-to-crankshaft arrangement which must be maintained when assembling these parts. If the piston is assembled in the bore

Timing marks for the LAV40 and HS40 engines

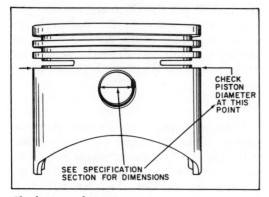

Check piston dimensions

180° out of position, it will cause immediate binding of the parts.

Piston Rings

Always replace the piston rings in sets. Ring gaps must be staggered. When using new rings, wipe the cylinder wall with fine emery cloth to deglaze the wall. Make sure the cylinder wall is thoroughly cleaned after deglazing. Check the ring gap by placing the ring squarely in the center of the area in which the rings travel and measuring the gap with a feeler gauge. Do not spread the rings too wide when assembling them to the pistons. Use a ring leader to install the rings on the piston.

The top compression ring has an inside chamfer. This chamfer must go UP. If the second ring has a chamfer, it must also face UP. If there is a notch on the outside diameter of the ring, it must face DOWN. Check the ring gap on the old ring to

determine if the ring should be replaced. Check the ring gap on the new ring to determine if the cylinder should be rebored to take oversize parts.

NOTE: *Make sure that the ring gap is measured with the ring fitted squarely in the worn part of the cylinder where the ring usually rides up and down on the piston.*

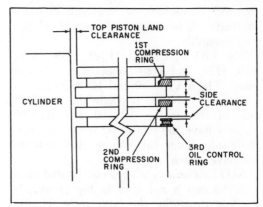

Ring arrangement and dimensions

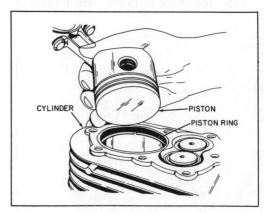

Squaring the ring in the bore

Checking the ring gap

Connecting Rods

Be sure that the match marks align when assembling the connecting rods to the crankshaft. Use new self-locking nuts. Whenever locking tabs are included, be sure that the tabs lock the nuts securely. NEVER try to straighten a bent crankshaft or connecting rod. Replace them if necessary. When replacing either the piston, rod, crankshaft, or camshaft, liberally lubricate all bearings with engine oil before assembly.

The LAV40, LAV50, HS40, HS50, V70, V80, H70, and H80 connecting rods are offset. The LAV40–50 and HS40-50 engine caps are fitted from opposite the camshaft side of the engine. V70–80 and H70–80 engines have the cap fitted from the camshaft side. Some early connecting rods do not have match marks.

NOTE: *Early type caps can be distorted if the cap is not held to the crank pin while threading the bolts tight. Undue force should not be used.*

Later rods have serrations which prevent

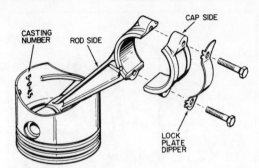

Connecting rod and piston assembly for the V80 and H80 engines

distortion during tightening. They also have match marks which must face out when assembling the rod. On the V80 and H80 engines, the piston and rod must fit so that the number inside the casting is on the rod side of the rod/cap combination. On the LAV50 and the HS50 engines, the piston must be fitted to the rod with the arrow on the top of the piston pointing to the right and the match marks on the rod facing you when the piston is pinned to the rod.

Camshaft

Before removing the camshaft, align the timing marks to relieve pressure on the valve lifters. Clean the camshaft in solvent, then blow the oil passages dry with compressed air. Replace the camshaft if it shows wear or evidence of scoring. Check the cam dimensions against those in the chart.

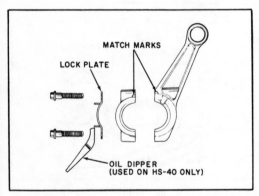

Connecting rod assembly for the LAV40 and HS40 engine

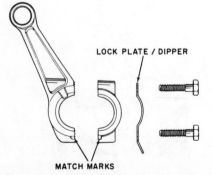

Connecting rod assembly for the V70, V80, H70, and H80 engines

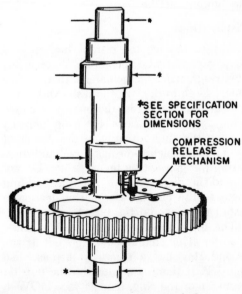

Check the dimensions of the camshaft

If the engine has a mechanical fuel pump, it may have to be removed to properly reinstall the camshaft. If the engine is equipped with the Insta-matic Ezee-Start Compression Release, and any of the parts have to be replaced due to wear or damage, the entire camshaft must be replaced. Be sure that the oil pump (if so equipped) barrel chamber is toward the fillet of the camshaft gear when assembled.

NOTE: *If a damaged gear is replaced, the crankshaft gear should also be replaced.*

Valve Springs

The valve springs should be replaced whenever an engine is overhauled. Check the free length of the springs. Comparing one spring with the other can be a quick check to notice any differences. If a difference is noticed, carefully measure the free length, compression length, and strength of each spring. See the specifications chart at the end of this section.

Valve Lifters

The stems of the valves serve as the lifters. On the 4 hp light frame models, the lifter stems are of different lengths. Because this engine is a cross port model, the shorter intake valve lifter goes nearest the mounting flange.

The valve lifters are identical on standard port engines. However once a wear pattern is established, they should not be interchanged.

Valve Grinding and Replacement

Valves and valve seats can be removed and reground with a minimum of engine disassembly.

Remove the valves as follows:

1. Raise the lower valve spring caps while holding the valve heads tightly against the valve seat to remove the valve spring retaining pin. This is best achieved by using a valve spring compressor. Remove the valves, springs, and caps from the crankcase.

2. Clean all parts with a solvent and remove all carbon from the valves.

3. Replace distorted or damaged valves. If the valves are in usable condition, grind the valve faces in a valve refacing machine and to the angle given in the specifications chart at the end of this section. Replace the valves if the faces are ground to less than $\frac{1}{32}$ in.

4. Whenever new or reground valves are installed, lap in the valves with lapping compound to insure an air-tight fit.

NOTE: *There are valves available with oversize stems.*

5. Valve grinding changes the valve lifter clearance. After grinding the valves check the valve lifter clearance as follows:

    a. Rotate the crankshaft until the piston is set at the TDC position of the compression stroke.

    b. Insert the valves in their guides and hold the valves firmly on their seats.

    c. Check for a clearance of 0.010 in. between each valve stem and valve lifter with a feeler gauge.

    d. Grind the valve stem in a valve resurfacing machine set to grind a perfectly square face with the proper clearance.

Replace the valves in the reverse order of removal. Make sure that the valve marked "EX" is in the exhaust valve position. Use a valve spring compressor to install the valve spring retainers.

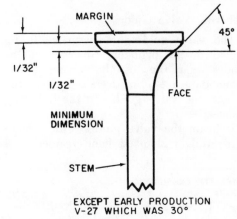

Dimensions of the valve face

Reboring the Cylinder

1. First, decide whether to rebore for 0.010 in. or 0.020 in.

2. Use any standard commercial hone of suitable size. Chuck the hone in the drill press with the spindle speed of about 600 rpm.

3. Start with coarse stones and center the cylinder under the press spindle. Lower the hone so the lower end of the

stones contact the lowest point in the cylinder bore.

4. Rotate the adjusting nut so that the stones touch the cylinder wall and then begin honing at the bottom of the cylinder. Move the hone up and down at a rate of 50 strokes a minute to avoid cutting ridges in the cylinder wall. Every fourth or fifth stroke, move the hone far enough to extend the stones 1 in. beyond the top and bottom of the cylinder bore.

5. Check the bore size and straightness every thirty or forty strokes. If the stones collect metal, clean them with a wire brush each time the hone is removed.

6. Hone with coarse stones until the cylinder bore is within 0.002 in. of the desired finish size. Replace the coarse stones with burnishing stones and continue until the bore is to within 0.0005 in. of the desired size.

7. Remove the burnishing stones and install finishing stones to polish the cylinder to the final size.

8. Clean the cylinder with solvent and dry it thoroughly.

9. Replace the piston and piston rings with the correct oversize parts.

Reboring Valve Guides

The valve guides are permanently installed in the cylinder. However, if the guides wear, they can be rebored to accommodate a $\frac{1}{32}$ in. oversize valve stem. Rebore the valve guides in the following manner:

1. Ream the valve guides with a standard straight shanked hand reamer or a low speed drill press. Refer to the specifications chart at the end of this section for the correct valve stem guide diameter.

2. Redrill the upper and lower valve spring caps to accommodate the oversize valve stem.

3. Reassemble the engine, installing valves with the correct oversize stems in the valve guides.

Regrinding Valve Seats

The valve seats need regrinding only if they are pitted or scored. If there are no

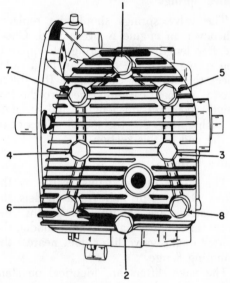

Cylinder head tightening sequence for all engines except 8 hp models

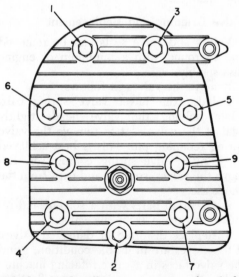

Cylinder head tightening sequence for 8 hp models

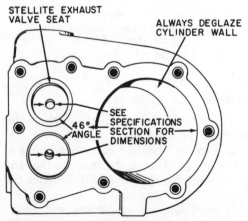

Check the dimensions of the valve seats, guides, and the cylinder

pits or scores, lapping in the valves will provide a proper valve seat. Valve seats are not replaceable. Regrind the valve seats as follows:

1. Use a grinding stone or a reseater set to provide the proper angle and seal face dimensions.

2. If the seat is over $\frac{3}{64}$ in. wide after grinding, use a 15° stone or cutter to narrow the face to the proper dimensions.

3. Inspect the seats to make sure that the cutter or stone has been held squarely to the valve seat and that the same dimension has been held around the entire circumference of the seat.

4. Lap the valves to the reground seats.

Bearing Service

Aluminum bearings must be cut out, using a reamer, and replaced with bronze bushings. Worn bronze bushings must be driven out before the new bushing can be installed.

On long life and cast iron engines with ball bearings on the power take-off (PTO) side of the crankshaft, the side cover containing the ball bearing must be removed and a cover with either a new bronze bushing or aluminum bearing must be substituted.

On the 2 to 3½ hp horizontal crankshaft engines, the snap-ring on the crankshaft (outside) must be removed in order to remove the cylinder cover. In order to remove the bearing, remove the crankshaft oil seal by prying it off and remove the snap-ring which holds the bearing in place. Reassemble in the reverse order of removal, using a new crankshaft oil seal.

The ball bearings on the 4 and 8 hp horizontal crankshaft engines are removed with a bearing splitter and a puller. It is installed after it is heated in hot oil. Refer to the "Briggs and Stratton" section for this procedure.

LUBRICATION

Barrel and Plunger Oil Pump System

This system is driven by an eccentric on the camshaft. Oil is drawn through the hollow camshaft from the oil sump on its intake stroke. The passage from the sump through the camshaft is aligned with the pump opening. As the camshaft continues rotation (pressure stroke), the plunger

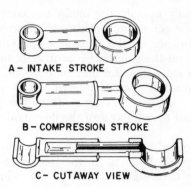

A – INTAKE STROKE

B – COMPRESSION STROKE

C – CUTAWAY VIEW

Operation of the barrel type oil pump

forces the oil out. The other port in the camshaft is aligned with the pump, and directs oil out of the top of the camshaft.

At the top of the camshaft, oil is forced through a crankshaft passage to the top main bearing groove which is aligned with the drilled crankshaft passage. Oil is directed through this passage to the crankshaft connecting rod journal and then spills from the connecting rod to lubricate the cylinder walls. Splash is used to lubricate the other parts of the engine.

A pressure relief port in the crankcase relieves excessive pressures when the oil viscosity is extremely heavy due to cold temperatures, or when the system is plugged or damaged. Normal pressure is 7 psi.

SERVICE

Remove the mounting flange or the cylinder cover, whichever is applicable. Remove the barrel and plunger assembly and separate the parts.

Clean the pump parts in solvent and inspect the pump plunger and barrel for rough spots or wear. If the pump plunger is scored or worn, replace the entire pump.

Before reassembling the pump parts, lubricate all of the parts in engine oil.

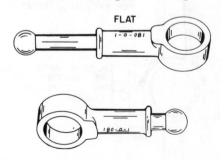

FLAT

FLAT MUST FACE OUT

Installation of the barrel type oil pump

Manually operate the pump to make sure the plunger slides freely in the barrel.

Lubricate all the parts and position the barrel on the camshaft eccentric. If the oil pump has a chamfer only on one side, that side must be placed toward the camshaft gear. The flat goes away from the gear, thus out to work against the flange oil pick-up hole.

Install the mounting flange. Be sure the plunger ball seats in the recess in the flange before fastening it to the cylinder.

Spray Mist Lubrication

Late model LAV40, LAV30, and LAV35 engines have a spray mist lubrication system. This system is the same as the barrel and plunger oil pump system except that (1) the pressure relief port is changed to a calibrated spray mist orifice and (2) the crankshaft is not rifle-drilled from the top main to the crank pin. Lubrication is sprayed to the narrow rod cap area through the spray mist hole.

Splash Lubrication

Some engines utilize the splash type lubrication system. The oil dipper, on some engines, is cast onto the lower connecting rod bearing cap. It is important that the proper parts are used to ensure the longest engine life.

Gear Type Oil Pump System

The gear type lubrication pump is a crankshaft driven, positive displacement pump. It pumps oil from the oil sump in the engine base to the camshaft, through the drilled camshaft passage to the top main bearing, through the drilled crankshaft, to the connecting rod journal on the crankshaft.

Spillage from the connecting rod lubricates the cylinder walls and normal splash lubricates the other internal working parts. There is a pressure relief valve in the system.

SERVICE

Disassemble the pump as follows: remove the screws, lockwashers, cover, gear, and displacement member.

Wash all of the parts in solvent. Inspect the oil pump drive gear and displacement member for worn or broken teeth, scoring, or other damage. Inspect the shaft hole in the drive gear for wear. Replace the entire pump if cracks, wear, or scoring is evident.

To replace the oil pump, position the oil pump displacement member and oil pump gear on the shaft, then flood all the parts with oil for priming during the initial starting of the engine.

The gasket provides clearance for the drive gear. With a feeler gauge, determine the clearance between the cover and the oil pump gear. The clearance desired is 0.006 to 0.007 in. Use gaskets, which are available in a variety of sizes, to obtain the correct clearance. Position the oil pump

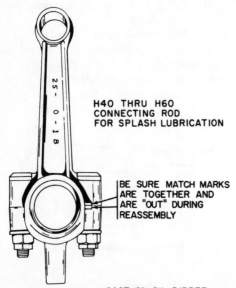

H40 THRU H60
CONNECTING ROD
FOR SPLASH LUBRICATION

BE SURE MATCH MARKS
ARE TOGETHER AND
ARE "OUT" DURING
REASSEMBLY

CAST ON OIL DIPPER

Splash type lubrication connecting rod

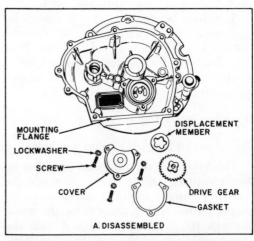

MOUNTING FLANGE

LOCKWASHER

SCREW

COVER

DISPLACEMENT MEMBER

DRIVE GEAR

GASKET

A. DISASSEMBLED

Disassembled view of the gear type lubrication system

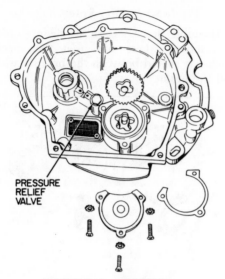

PRESSURE
RELIEF
VALVE

B. PARTIALLY ASSEMBLED

Partially assembled view of the gear type lubrication system

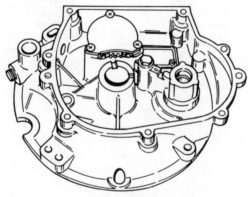

C. ASSEMBLED

View of the assembled gear type lubrication system

cover and secure it with the screws and lockwashers.

REWIND AND WIND-UP STARTERS

Disassembly

REWIND FRICTION SHOE STARTER

Relieve spring tension before disassembling the starter. Hold the rotor with your thumb after guiding the rope in the rotor notch then slowly let the spring unwind. Note the position of the friction shoes for reassembly.

CAUTION: *Be careful when removing or installing the spring; be sure it is wound in the correct direction.*

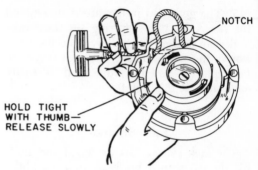

NOTCH

HOLD TIGHT
WITH THUMB—
RELEASE SLOWLY

Releasing the spring tension

REWIND DOG STARTER

Relieve the spring tension before disassembling. Hold the pulley firmly after guiding the rope into the pulley notch, then slowly let the spring unwind. Note the position of the dog for correct reassembly.

CAUTION: *Be careful when removing or installing the spring. Be sure it is wound in the correct direction.*

SIDE MOUNTED STARTER

1. Remove the handle.
2. Relieve the spring tension by allowing the starter spring cover to slide slowly through your fingers as the rope end slips past the rope clip. Complete removal of the starter is not always necessary if it is easily accessible.

NOTE: *It may be necessary to remove the shrouding.*

3. Remove the starter spring cover by loosensing the two small screws. This exposes the rewind spring which can easily be replaced at this point if necessary.

CAUTION: *Always remove the spring tension before removing the cover.*

4. A temporary spring retainer makes spring replacement easier. After removing the old spring, reinsert the replacement spring in the same relative position. Lay the spring retainer over the pulley spring receptacle and push the spring out of the retainer into position. Discard the retainer.

5. To further disassemble the starter, remove the center hub screw and spring hub. This will allow the pulley and gear to be removed as an assembly.

6. The brake spring may be removed and replaced. Check the spring for operative condition; it must cling snugly into the groove receptacle in the gear.

7. The gear and pulley are disassembled by removing the snap-ring on the brake and gear side of the pulley. Note the location of the steel washer. The gear may now be replaced and the rope end is easily accessible at this point.

KICK STARTER

The kick starter was designed and built for the 2.5 thru 5 hp Tecumseh mini-bike engines.

1. Remove the crankarm assembly by removing the screw. If necessary, mark the arm and starter for proper reinstallation when repairs have been completed.

2. Remove the four screws which hold the cover assembly to the starter housing. Hold the dog plate and pin assembly so that spring tension will not be released when the cover is removed from the housing. Remove the cover from the housing.

3. When the cover has cleared the housing, gradually release the spring tension by allowing the dog plate and pin, dog, and cam to unwind in the direction opposite the arrow.

4. Remove the retainer ring. Remove the shaft, clutch plate and liner, gear and hex nut as an assembly. Remove the thrust washer, spring, and keeper.

5. Remove the center screw to gain access to the individual parts assembled to the dog plate and pin and shaft. Note the location of the dog, the dog return spring, and location of the brake on the plate and on the dog cam.

6. Remove the hex nut (left-hand thread) from the shaft for servicing the gear, clutch liner, or the clutch plate.

7. Examine the parts for wear or damage and replace parts as necessary.

Assembly

REWIND FRICTION SHOE STARTER

1. Assemble the starter in the reverse order of disassembly, using a new retaining ring and new centering pin where it is applicable.

2. Place the rope in the rotor notch, then wind about six times. Slowly let the recoil wind up the rope. Check it for proper tension.

REWIND DOG STARTER

Assemble the starter in the reverse order of removal. Make sure that the spacer washer is seated between the brake washer and the brake.

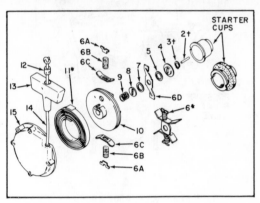

Exploded view of a recoil friction shoe starter

2.	Centering pin	7.	Washer
3.	Retaining ring	8.	Washer
4.	Retainer	9.	Spring
5.	Washer	10.	Sheave
6D.	Pawl lockout	11.	Spring
6C.	Pawl	12.	Retainer insert
6B.	Spring	13.	Handle
6A.	Pawl retainer	14.	Rope
6.	Pawls assembly (friction shoes)	15.	Housing

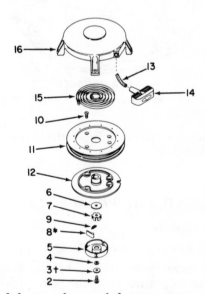

Exploded view of a recoil dog type starter

2.	Bolt	9.	Key
3.	Spacer	10.	Screw
4.	Washer	11.	Sheave
5.	Starter dog retainer plate	12.	Retainer plate
6.	Washer	13.	Rope
7.	Dog catch	14.	Handle
8.	Starter dog	15.	Housing

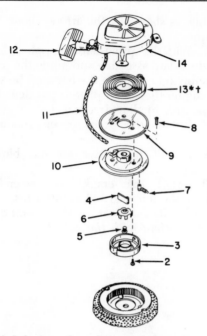

Exploded view of another recoil dog type starter

2. Screw
3. Starter dog retainer plate
4. Starter dog
5. Centering pin
6. Dog catch
7. Dog return spring
8. Screw

9. Sheave
10. Retainer plate
11. Rope
12. Handle
13. Spring
14. Housing

LATER MODELS HAVE
CENTERING PIN AND
SLEEVE FOR ALIGNING
STARTER TO CRANKSHAFT

GREASE LIGHTLY

LOCK TABS

*DISASSEMBLY—
REFER TO TEXT
†ASSEMBLY —
REFER TO TEXT

Rewind starter, dog type #590408, exploded view

1. Rope
2. Centering pin
3. Screw
4. Starter dog retainer
5. Brake spring
6. Starter dog

7. Dog return spring
8. Sheave
9. Spring and spring keeper
10. Handle and retainer
11. Housing

SIDE MOUNTED STARTER

Reassemble the starter in the reverse order of disassembly.

Lubricate the center shaft and the edges of the rewind spring with light grease.

LIGHTLY GREASE

HOOK CENTER END INTO SPRING HOLE

REWIND SPRING

SPRING HUB

THREE (3) MOUNTING SCREWS

OUTER SPRING END FASTENING POST

ROPE CLIP

NOTE: RELIEVE TENSION BEFORE REMOVING COVER

HUB SCREW MUST BE TIGHT FOR PROPER FUNCTION. REMOVE TO DISASSEMBLE

Side mounted starter

KICK STARTER

1. Install the clutch plate, clutch liner, gear (flat side toward liner), thin flat washer, thick flat washer, belleville washer (crown toward hex nut), and the hex nut (flat side against belleville washer) in that order on the shaft. Tighten the nut to 40 ft lbs maximum.

2. Install the spring and keeper assembly on the hex nut with the spring end in the notch on the nut.

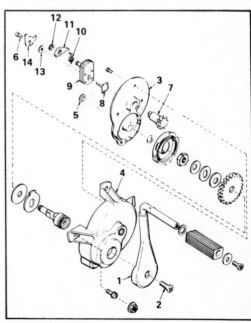

Exploded view of the kick starter

3. Place the thrust washer on the shaft and install the spring and keeper and gear assembly in the cover.

NOTE: *The locators on the keeper must fit in the cover. Install the clip on the shaft opposite the spring side of the cover.*

4. Install the shaft through the cover. Install the brake spring on the cover.

5. Install the dog plate, dog spring, clip, washer, dog cam, and center screw on the shaft. Tighten the screw to 60 ft lbs.

6. Holding the cover with the gear assemblies in place, wind the dog plate assembly, as shown by the arrow, about 2 or 3 turns.

7. With the spring tension applied, hold the dog plate assembly and guide the plate assembly into place in the housing. Be sure the cog on the clutch plate is positioned in the housing to allow free travel.

8. Install the four cover retaining screws.

9. Install the starter on the blower housing.

10. Install the crank arm assembly. Tighten the screw to 120 to 140 ft lbs.

NOTE: *Be sure that the starter arm does not interfere with the vehicle.*

FOUR-CYCLE VERTICAL CRANKSHAFT ENGINES

LV, LAV & LVC SERIES				LVC, LVR, LVT, LVTC, LVTR, LAVC, LAVR, LAVT, LAVTC & LAVTR SERIES				SLV, VA, V & VH SERIES		VC, VT, VX, VCX & VXT SERIES			
Model No.	See Column	Model No.	See Column	Model No.	See Column	Model No.	See Column	Model No.	See Column	Model No.	See Column	Model No.	See Column
LV22A	B	LAV25G	B	LVC22L	B	LVR30M	C	SLV	A	VC35B	E	VXT45C	G
LV22B	B	LAV25H	B	LVC22M	B	LVR30N	C	VA	B	VC40B	E	VXT55A	F
LV22C	B	LAV25J	B	LVC22N	B	LVT30A	C	V-17	A	VC45A	E	VXT55B	F
LV22D	B	LAV25K	B	LVR22L	B	LVT30B	C	V-17B	A	VC45B	E	VXT55C	F
LV22E	B	LAV25L	B	LVR22M	B	LVT30C	C	V-17D	A	VC45C	E		
LV22G	B	LAV25M	B	LVR22N	B	LVTC30A	C	V20	A	VC45D	F		
LV22H	B	LAV30	H	LVT22A	B	LVTC30B	C	V20B	A	VC55A	F		
LV22J	B	LAV30A	C	LVT22B	B	LVTC30C	C	V20D	A	VC55B	F		
LV22K	B	LAV30B	C	LVT22C	B	LVTR30A	C	V20F	A	VC55C	F		
LV22L	B	LAV30D	C	LVTC22A	B	LVTR30B	C	V20G	A	VC55D	F		
LV22M	B	LAV30E	C	LVTC22B	B	LVTR30C	C	V22	B	VT45A	G		
LV22N	B	LAV30G	C	LVTC22C	B	LAVC30L	C	V22B	B	VT45C	G		
LV25A	B	LAV30H	C	LVTR22A	B	LAVC30M	C	V22D	B	VT55A	F		
LV25B	B	LAV30J	C	LVTR22B	B	LAVR30L	C	V22H	B	VT55B	F		
LV25C	B	LAV30K	C	LVTR22C	B	LAVR30M	C	V25	B	VT55C	F		
LV25D	B	LAV30L	C	LAVC22L	B	LAVT30A	C	V25B	B	VX20F	A		
LV25E	B	LAV30M	C	LAVC22M	B	LAVT30B	C	V25C	B	VX20G	A		
LV25G	B	LAV35	D	LAVR22L	B	LAVT30C	C	V25D	B	VX22H	B		
LV25H	B	LAV35A	D	LAVR22M	B	LAVTC30A	C	V25E	B	VX25F	B		
LV25J	B	LAV35B	D	LAVT22A	B	LAVTC30B	C	V25F	B	VX25G	B		
LV25K	B	LAV35D	D	LAVT22B	B	LAVTC30C	C	V25G	B	VX25H	B		
LV25L	B	LAV35E	D	LAVT22C	B	LAVTR30A	C	V25H	B	VX27F	B		
LV25M	B	LAV35F	D	LAVTC22A	B	LAVTR30B	C	V27B	B	VX27G	B		
LV25N	B	LAV35G	D	LAVTC22B	B	LAVTR30C	C	V27D	B	VX27H	B		
LV30A	C	LAV35H	D	LAVTC22C	B	LVC35L	D	V27F	B	VX30F	C		
LV30B	C	LAV35J	D	LAVTR22A	B	LVC35M	D	V27G	B	VX30G	C		
LV30D	C	LAV35K	D	LAVTR22B	B	LVC35N	D	V27H	B	VX30H	C		
LV30E	C	LAV35L	D	LAVTR22C	B	LVR35L	D	V30F	C	VX32A	D		
LV30G	C	LAV35M	D	LVC25L	B	LVR35M	D	V30G	C	VX35A	E		
LV30H	C	LAV40	J	LVC25M	B	LVR35N	D	V30H	C	VX35B	E		
LV30J	C	LCV22A	B	LVC25N	B	LVT35A	D	V32A	D	VX40A	E		
LV30K	C	LCV22B	B	LVR25L	B	LVT35B	D	V35A	E	VX40B	E		
LV30L	C	LCV22C	B	LVR25M	B	LVT35C	D	V35B	E	VX45A	E		
LV30M	C	LCV22D	B	LVR25N	B	LVTC35A	D	V40	E	VX45B	E		
LV30N	C	LCV25A	B	LVT25A	B	LVTC35B	D	V40A	E	VX45C	E		
LV35	I	LCV25B	B	LVT25B	B	LVTC35C	D	V40B	E	VX45D	F		
LV35A	D	LCV25C	B	LVT25C	B	LVTR35A	D	VH40	E	VX55A	F		
LV35B	D	LCV25D	B	LVTC25A	B	LVTR35B	D	V45A	E	VX55B	F		
LV35D	D	LCV30A	C	LVTC25B	B	LVTR35C	D	V45B	E	VX55C	F		
LV35E	D	LCV30B	C	LVTC25C	B	LAVC35L	D	V45C	E	VX55D	F		
LV35F	D	LCV30D	C	LVTR25A	B	LAVC35M	D	V45D	F	VCX35B	E		
LV35G	D	LCV35A	D	LVTR25B	B	LAVR35L	D	V50	G	VCX40B	E		
LV35H	D	LCV35B	D	LVTR25C	B	LAVR35M	D	VH50	G	VCX45A	E		
LV35J	D	LCV35D	D	LAVC25L	B	LAVT35A	D	V55A	F	VCX45B	E		
LV35K	D			LAVC25M	B	LAVT35B	D	V55B	F	VCX45C	E		
LV35L	D			LAVR25L	B	LAVT35C	D	V55C	F	VCX45D	F		
LV35M	D			LAVR25M	B	LAVTC35A	D	V55D	F	VCX55A	F		
LV35N	D			LAVT25A	B	LAVTC35B	D	V60	F	VCX55B	F		
LAV22	B			LAVT25B	B	LAVTC35C	D	VH60	F	VCX55C	F		
LAV22A	B			LAVT25C	B	LAVTR35A	D	V70	K	VCX55D	F		
LAV22B	B			LAVTC25A	B	LAVTR35B	D	VH70	K	VXT45A	G		
LAV22C	B			LAVTC25B	B	LAVTR35C	D						
LAV22D	B			LAVTC25C	B								
LAV22E	B			LAVTR25A	B								
LAV22G	B			LAVTR25B	B								
LAV22H	B			LAVTR25C	B								
LAV22J	B			LVC30L	C								
LAV22K	B			LVC30M	C								
LAV22L	B			LVC30N	C								
LAV22M	B			LVR30L	C								
LAV25	H												
LAV25A	B												
LAV25B	B												
LAV25C	B												
LAV25D	B												
LAV25E	B												

Model number-to-column letter cross reference chart

FOUR-CYCLE HORIZONTAL CRANKSHAFT ENGINES

H, HH & HT SERIES

Model No.	See Column	Model No.	See Column
H20A	A	HT35A	D
H20B	A	HT35B	I
H20C	A	H40	E
		HS40	J
H22	B	H45A	G
H22D	B	H45B	G
H22E	B	H45D	G
H22F	B	HT45A	G
H22H	B	HT45C	G
H22J	B		
H22K	B	H50	G
H22L	B		
H22M	B	H55	G
H22N	B	H55A	G
H22P	B	H55B	G
H22R	B	H55D	F
HT22A	B	HT55A	F
HT22B	B	HT55B	F
		HT55C	F
H25	H		
H25A	B	H60	F
H25B	B		
H25C	B	HH40	E
H25D	B		
H25E	B	HH50	G
H25F	B		
H25H	B	HH60	F
H25J	B		
H25K	B	H70	K
H25L	B		
H25M	B	HH70	K
H25N	B		
H25P	B		
H25R	B		
HT25A	B		
HT25B	H		
H30	H		
H30A	C		
H30B	C		
H30C	C		
H30D	C		
H30E	C		
H30H	C		
H30J	C		
H30K	C		
H30L	C		
H30M	C		
H30N	C		
H30P	C		
H30R	C		
HT30A	C		
HT30B	H		
H35	I		
H35D	D		
H35E	D		
H35H	D		
H35K	D		
H35M	D		
H35P	D		
H35R	D		

HA, HB, HTB & HTC SERIES

Model No.	See Column	Model No.	See Column
HA22A	A	HTB35B	D
HA22B	A	HTC35A	D
HA22C	A	HTC35B	D
HB22D	B	HB45A	G
HB22E	B	HB45B	G
HB22F	B	HB45D	G
HB22H	B	HTB45A	G
HB22J	B	HTB45C	G
HB22K	B		
HB22L	B	HB55A	G
HB22M	B	HB55B	G
HB22N	B	HB55D	F
HB22P	B	HTB55A	F
HTB22A	B	HTB55B	F
HTB22B	B	HTB55C	F
HTC22A	B		
HTC22B	B		
HB25A	B		
HB25B	B		
HB25C	B		
HB25D	B		
HB25E	B		
HB25F	B		
HB25H	B		
HB25J	B		
HB25K	B		
HB25L	B		
HB25M	B		
HB25N	B		
HB25P	B		
HTB25A	B		
HTB25B	B		
HTC25A	B		
HTC25B	B		
HB30A	C		
HB30B	C		
HB30D	C		
HB30E	C		
HB30F	C		
HB30H	C		
HB30J	C		
HB30K	C		
HB30L	C		
HB30M	C		
HB30N	C		
HB30P	C		
HTB30A	C		
HTB30B	C		
HTC30A	C		
HTC30B	C		
HB35	I		
HB35D	D		
HB35E	D		
HB35H	D		
HB35K	D		
HB35M	D		
HB35P	D		
HTB35A	D		

HR & HTR SERIES

Model No.	See Column	Model No.	See Column
HR20A	A	HR55A	G
HR20B	A	HR55B	G
HR20C	A	HR55D	F
		HTR55A	F
HR22D	B	HTR55B	F
HR22E	B	HTR55C	F
HR22F	B		
HR22H	B		
HR22J	B		
HR22K	B		
HR22L	B		
HR22M	B		
HR22N	B		
HR22P	B		
HTR22A	B		
HTR22B	B		
HR25A	B		
HR25B	B		
HR25C	B		
HR25D	B		
HR25E	B		
HR25F	B		
HR25H	B		
HR25J	B		
HR25K	B		
HR25L	B		
HR25M	B		
HR25N	B		
HR25P	B		
HTR25A	B		
HTR25B	B		
HR30A	C		
HR30B	C		
HR30C	C		
HR30D	C		
HR30E	C		
HR30H	C		
HR30J	C		
HR30K	C		
HR30L	C		
HR30M	C		
HR30N	C		
HR30P	C		
HTR30A	C		
HTR30B	C		
HR35D	D		
HR35E	D		
HR35H	D		
HR35K	D		
HR35M	D		
HR35P	D		
HTR35A	D		
HTR35B	D		
HR45A	E		
HR45B	E		
HR45D	G		
HTR45A	G		
HTR45C	G		

Model number-to-column letter cross reference chart

VERTICAL CRAFTSMAN ENGINES

Craftsman Engine Models	See Column	Craftsman Engine Models	See Column	Craftsman Engine Models	See Column	Craftsman Engine Models	See Column
143.V20A-1-1-1WA	A	143.30251	B	143.50021	A	143.60231	C
143.V25A-5-1-1WA	B	143.30350	C	143.50025	B	143.60240	E
143.V27A-3-1-1WA	B	143.30351	C	143.50026	B	143.60325	B
143.09300	B	143.31600	F	143.50030	C	143.60326	B
143.09301	B	143.31601	F	143.50031	C	143.60327	B
143.09302	B	143.36250	A	143.50040	E	143.60328	B
143.10250	B	143.36251	A	143.50045	E	143.60330	B
143.10251	B	143.36252	B	143.50125	B	143.60331	B
143.10300	B	143.36253	B	143.50126	B	143.60340	E
143.10301	B	143.36254	B	143.50130	C	143.60351	C
143.11300	B	143.36255	B	143.50131	C	143.60401	D
143.11301	B	143.39250	B	143.50230	C	143.62200	B
143.11302	B	143.39251	B	143.50231	C	143.62201	B
143.11350	B	143.40250	B	143.50250	B	143.62700	B
143.11351	B	143.40251	B	143.50251	B	143.62701	B
143.11352	B	143.40300	B	143.50300	C	143.62702	B
143.12300	B	143.40301	B	143.50301	C	143.62703	B
143.12301	B	143.40350	C	143.50400	D	143.62704	B
143.12302	B	143.40351	C	143.50401	D	143.62705	B
143.12303	B	143.40500	E	143.50402	D	143.63200	C
143.12304	B	143.40501	E	143.50403	D	143.63201	C
143.12350	B	143.40502	E	143.52200	B	143.63202	C
143.12351	B	143.40503	E	143.52201	B	143.63203	C
143.13250	B	143.40600	F	143.52700	B	143.63204	C
143.13251	B	143.41300	B	143.52701	B	143.63205	C
143.13300	B	143.41301	B	143.52702	B	143.64000	E
143.13301	B	143.41302	B	143.52703	B	143.64500	E
143.14350	C	143.41350	C	143.53001	C	143.65250	B
143.14351	C	143.41351	C	143.53200	C	143.65251	B
143.15300	B	143.41352	C	143.53201	C	143.66250	B
143.15301	B	143.42200	B	143.53300	C	143.66251	B
143.16350	C	143.42201	B	143.53301	C	143.66500	E
143.16351	C	143.42202	B	143.54500	E	143.66501	E
143.17350	C	143.42203	B	143.54502	E	143.67250	B
143.17351	C	143.42204	B	143.55250	A	143.67251	B
143.18350	C	143.42205	B	143.55251	A	143.71250	B
143.18351	C	143.42500	B	143.55300	C	143.71251	B
143.19400	D	143.42501	B	143.56250	A	143.75250	B
143.19401	D	143.42700	B	143.56251	A	143.75251	B
143.20100	A	143.42701	B	143.56252	A	143.76250	B
143.20400	D	143.43004	C	143.56253	A	143.76251	B
143.20401	D	143.43200	C	143.58250	A	143.76252	B
143.20500	E	143.43201	C	143.58251	A	143.82000	A
143.20501	E	143.43202	C	143.59250	B	143.82001	A
143.20502	E	143.43203	C	143.59251	B	143.82002	A
143.20503	E	143.43204	C	143.60020	A	143.82003	A
143.23250	A	143.43205	C	143.60021	A	143.82004	A
143.23251	A	143.43250	B	143.60025	B	143.82005	A
143.24250	B	143.43251	B	143.60026	B	143.82006	A
143.24251	B	143.43300	C	143.60030	C	143.82007	A
143.25100	B	143.43301	C	143.60031	C	143.83000	C
143.25250	B	143.43500	E	143.60040	E	143.83001	C
143.25251	B	143.43501	E	143.60125	B	143.83250	B
143.27100	B	143.43700	D	143.60126	B	143.83251	B
143.27200	A	143.43701	D	143.60130	C	143.83252	B
143.30250	B	143.44000	E	143.60131	C	143.84300	B
		143.44400	D	143.60140	E	143.84301	B
		143.44401	D	143.60225	B	143.84302	B
		143.50020	A	143.60226	B	143.86400	D

Model number-to-column letter cross reference chart

VERTICAL CRAFTSMAN ENGINES (Cont.)

Craftsman Engine Models	See Column	Craftsman Engine Models	See Column	Craftsman Engine Models	See Column	Craftsman Engine Models	See Column
143.86401	D	143.102162	B	143.105031	F	143.122321	C
143.86402	D	143.102170	C	143.105040	F	143.122322	C
143.90020	A	143.102171	C	143.105041	F	143.123021	C
143.90021	A	143.102172	C	143.105050	F	143.123022	C
143.91250	B	143.102190	C	143.105051	F	143.123031	C
143.91251	B	143.102191	C	143.105060	F	143.123041	C
143.93000	C	143.102200	C	143.105061	F	143.123051	C
143.97250	B	143.102201	C	143.105070	F	143.123052	C
143.97251	B	143.102202	C	143.105071	F	143.123071	C
143.101010	C	143.102210	C	143.105090	F	143.123091	C
143.101011	C	143.102211	C	143.105091	F	143.123092	C
143.101012	C	143.102212	C	143.105100	F	143.124011	D
143.101020	C	143.102220	C	143.105101	F	143.124021	D
143.101021	C	143.102221	C	143.105110	F	143.124031	D
143.101022	C	143.102222	C	143.105120	F	143.124041	D
143.101030	C	143.102232	C	143.105121	F	143.124051	D
143.101031	C	143.102231	C	143.106010	C	143.124061	D
143.101032	C	143.102232	C	143.106011	C	143.124071	D
143.102010	C	143.102240	C	143.106012	C	143.125011	G
143.102011	C	143.102241	C	143.106020	C	143.125021	G
143.102012	C	143.102242	C	143.106021	C	143.125031	G
143.102020	C	143.102311	C	143.106030	C	143.125041	G
143.102021	C	143.103010	C	143.106031	C	143.125051	G
143.102022	C	143.103011	C	143.122011	C	143.125061	G
143.102030	C	143.103012	C	143.122012	C	143.126011	F
143.102031	C	143.103020	C	143.122021	C	143.126021	F
143.102040	C	143.103021	C	143.122022	C	143.126031	F
143.102041	C	143.103050	C	143.122031	C	143.126041	F
143.102042	C	143.103051	C	143.122032	C	143.126051	F
143.102050	C	143.103052	C	143.122041	C	143.126061	F
143.102051	C	143.103060	C	143.122042	C	143.131022	H
143.102060	C	143.103061	C	143.122051	C	143.131032	H
143.102061	C	143.104010	D	143.122052	C	143.131042	H
143.102062	C	143.104011	D	143.122061	C	143.131052	H
143.102070	C	143.104020	D	143.122071	C	143.131062	H
143.102071	C	143.104021	D	143.122081	C	143.131072	H
143.102072	C	143.104030	D	143.122082	C	143.131082	H
143.102080	C	143.104031	D	143.122091	C	143.131092	H
143.102081	C	143.104040	D	143.122092	C	143.131102	H
143.102090	C	143.104041	D	143.122101	B	143.131112	B
143.102091	C	143.104050	D	143.122102	B	143.131122	H
143.102100	C	143.104051	D	143.122201	C	143.131132	H
143.102101	C	143.104060	D	143.122202	C	143.131142	H
143.102102	C	143.104061	D	143.122211	C	143.131152	H
143.102110	C	143.104080	D	143.122212	C	143.131162	H
143.102111	C	143.104081	D	143.122221	C	143.131172	H
143.102112	C	143.104090	D	143.122222	C	143.131182	H
143.102120	C	143.104091	D	143.122231	C	143.133012	H
143.102121	C	143.104100	D	143.122232	C	143.133022	H
143.102130	C	143.104101	D	143.122251	C	143.133032	H
143.102131	C	143.104110	D	143.122252	C	143.133042	H
143.102132	C	143.104111	D	143.122261	C	143.133052	H
143.102140	C	143.104120	D	143.122262	C	143.134012	I
143.102141	C	143.104121	D	143.122271	C	143.134022	I
143.102150	C	143.105010	F	143.122272	C	143.134032	I
143.102151	C	143.105011	F	143.122281	C	143.134042	I
143.102152	C	143.105020	F	143.122282	C	143.134052	I
143.102160	B	143.105021	F	143.122291	C	143.135012	G
143.102161	B	143.105030	F	143.122311	C	143.135022	G

Model number-to-column letter cross reference chart

VERTICAL CRAFTSMAN ENGINES (Cont.)

Craftsman Engine Models	See Column	Craftsman Engine Models	See Column	Craftsman Engine Models	See Column	Craftsman Engine Models	See Column
143.135042	G	143.145022	G	143.161092	H	143.171062	H
143.135052	G	143.145032	G	143.161102	H	143.171072	H
143.135062	G	143.145042	G	143.161112	H	143.171082	H
143.135072	G	143.145052	G	143.161132	H	143.171092	H
143.135082	G	143.145062	G	143.161142	H	143.171102	H
143.135092	G	143.145072	G	143.161152	H	143.171132	H
143.135102	G	143.146012	F	143.161162	H	143.171142	H
143.135112	G	143.146022	F	143.161172	H	143.171152	H
143.136012	F	143.147012	E	143.161182	H	143.171162	H
143.136032	F	143.147022	E	143.161192	H	143.171172	H
143.136042	F	143.147032	E	143.161202	H	143.171202	D
143.136052	F	143.151012	H	143.161212	H	143.171212	H
143.137012	E	143.151022	H	143.161222	H	143.171232	H
143.137032	E	143.151032	H	143.161232	H	143.171242	H
143.141012	H	143.151042	H	143.161242	H	143.171252	H
143.141022	H	143.151052	H	143.161262	H	143.171262	H
143.141032	H	143.151072	H	143.163012	H	143.171302	H
143.141042	H	143.151082	H	143.163022	H	143.171312	H
143.141052	H	143.151092	H	143.163032	H	143.171322	H
143.141062	H	143.151102	H	143.163042	H	143.171332	H
143.141072	H	143.151103	H	143.163052	H	143.173012	H
143.141082	H	143.151104	H	143.163062	H	143.173042	H
143.141092	H	143.151105	H	143.164012	D	143.174012	D
143.141102	H	143.151142	H	143.164022	D	143.174022	D
143.141112	H	143.153012	H	143.164032	D	143.174032	D
143.141122	H	143.153022	H	143.164042	D	143.174042	D
143.141132	H	143.153032	H	143.164052	D	143.174052	D
143.141142	H	143.154012	D	143.164062	D	143.174062	D
143.141152	H	143.154022	D	143.164072	D	143.174072	D
143.141162	H	143.154032	D	143.164082	D	143.174082	D
143.141172	H	143.154042	D	143.164102	D	143.174092	D
143.141182	H	143.154052	D	143.164112	D	143.174102	D
143.141192	H	143.154062	D	143.164122	D	143.174132	D
143.141202	H	143.154072	D	143.164132	D	143.174142	D
143.141212	H	143.154082	D	143.164142	D	143.174152	D
143.141222	H	143.154092	D	143.164152	D	143.174162	D
143.141232	H	143.154102	D	143.164162	D	143.174172	D
143.141242	H	143.154112	D	143.164172	D	143.174182	D
143.141252	H	143.154132	D	143.164182	D	143.174192	D
143.141262	H	143.154142	D	143.164202	D	143.174232	D
143.141272	H	143.155012	G	143.165012	G	143.174242	D
143.141282	H	143.155022	G	143.165022	G	143.174252	D
143.143012	H	143.155032	G	143.165032	G	143.174272	D
143.143022	H	143.155042	G	143.165042	G	143.174292	D
143.143032	H	143.155052	G	143.165052	G	143.175012	G
143.144012	I	143.155062	G	143.166012	F	143.175022	G
143.144022	I	143.156012	F	143.166022	F	143.175032	G
143.144032	I	143.156022	F	143.166032	F	143.175042	G
143.144042	I	143.157012	E	143.166042	F	143.175052	G
143.144052	I	143.157022	E	143.166052	F	143.175062	G
143.144062	I	143.157032	E	143.167012	E	143.175072	G
143.144072	I	143.161012	H	143.167022	E	143.176012	F
143.144082	I	143.161022	H	143.167032	E	143.176022	F
143.144092	I	143.161032	H	143.167042	E	143.176032	F
143.144102	I	143.161042	H	143.171012	H	143.176042	F
143.144112	I	143.161052	H	143.171022	H	143.176052	F
143.144122	I	143.161062	H	143.171032	H	143.176062	F
143.144132	I	143.161072	H	143.171042	H	143.176072	F
143.145012	G	143.161082	H	143.171052	H	143.176082	F

Model number-to-column letter cross reference chart

VERTICAL CRAFTSMAN ENGINES (Cont.)

Craftsman Engine Models	See Column	Craftsman Engine Models	See Column	Craftsman Engine Models	See Column	Craftsman Engine Models	See Column
143.176092	F	143.186122	F	143.204142	D		
143.177012	E	143.187022	J	143.204162	D		
143.177022	E	143.187042	J	143.204172	D		
143.177032	E	143.187052	J	143.204182	D		
143.177042	E	143.187062	J	143.204192	D		
143.177062	E	143.187072	J	143.204202	L		
143.177072	E	143.187082	J	143.205022	G		
143.181042	H	143.187094	J	143.206012	F		
143.181052	H	143.187102	J	143.206022	K		
143.181062	H	143.191012	H	143.206032	F		
143.181082	H	143.191022	H	143.207012	J		
143.181092	H	143.191032	H	143.207022	J		
143.181102	H	143.191042	H	143.207032	J		
143.181112	H	143.191052	H	143.207052	J		
143.181122	H	143.194012	D	143.207062	M		
143.181132	H	143.194022	D	143.207072	J		
143.183022	H	143.194032	D	143.207082	M		
143.183042	H	143.194042	D				
143.184012	D	143.194052	D				
143.184052	D	143.194062	L				
143.184082	D	143.194072	D				
143.184092	D	143.194082	D				
143.184102	D	143.194092	D				
143.184112	D	143.194102	L				
143.184122	D	143.194112	D				
143.184132	D	143.194122	D				
143.184142	D	143.194132	D				
143.184152	D	143.194142	D				
143.184162	D	143.195012	G				
143.184172	D	143.195022	G				
143.184182	D	143.196012	F				
143.184192	D	143.196022	F				
143.184202	D	143.196032	F				
143.184212	D	143.196042	K				
143.184232	L	143.196052	K				
143.184242	L	143.196062	K				
143.184252	L	143.196072	K				
143.184262	D	143.196082	F				
143.184272	D	143.197012	J				
143.184282	D	143.197022	M				
143.184292	D	143.197032	M				
143.184302	D	143.197042	J				
143.184402	D	143.197052	J				
143.185012	G	143.197062	J				
143.185022	G	143.197072	J				
143.185032	G	143.197082	M				
143.185042	G	143.201032	H				
143.185052	G	143.201042	H				
143.186012	F	143.203012	H				
143.186022	K	143.204022	L				
143.186032	K	143.204032	D				
143.186042	K	143.204042	D				
143.186052	F	143.204052	D				
143.186062	K	143.204062	L				
143.186072	F	143.204072	D				
143.186082	F	143.204082	D				
143.186092	F	143.204092	D				
143.186102	F	143.204102	L				
143.186112	F	143.204132	L				

Model number-to-column letter cross reference chart

HORIZONTAL CRAFTSMAN ENGINES (cont.)

Craftsman Engine Models	See Column	Craftsman Engine Models	See Column	Craftsman Engine Models	See Column	Craftsman Engine Models	See Column
143.501011	B	143.526031	F	143.541292	H	143.556112	F
143.501021	B	143.531022	B	143.541302	H	143.556122	F
143.501041	B	143.531032	B	143.544012	I	143.556132	F
143.501051	B	143.531042	B	143.544022	I	143.556142	F
143.501061	B	143.531052	H	143.544032	I	143.556152	F
143.501071	B	143.531062	B	143.544042	I	143.556162	F
143.501081	B	143.531082	H	143.545012	G	143.556172	F
143.501090	B	143.531092	B	143.545022	G	143.556182	F
143.501091	B	143.531112	B	143.545032	G	143.556192	F
143.501111	B	143.531122	H	143.545042	G	143.556202	F
143.501121	B	143.531132	H	143.546012	F	143.556212	F
143.501131	B	143.531142	B	143.546022	F	143.556222	F
143.501141	B	143.531152	H	143.546032	F	143.556232	F
143.501151	B	143.531162	B	143.546042	F	143.556242	F
143.501161	B	143.531172	H	143.546052	F	143.556252	F
143.501170	B	143.531182	H	143.546062	F	143.556262	F
143.501171	B	143.534012	I	143.546072	F	143.556272	F
143.501181	B	143.534022	I	143.546082	F	143.556282	F
143.501190	B	143.534032	I	143.546092	F	143.557012	E
143.501191	B	143.534042	I	143.546102	F	143.557022	E
143.501201	B	143.534052	I	143.546112	F	143.557032	E
143.501211	B	143.534062	I	143.546122	F	143.557042	E
143.501221	B	143.534072	I	143.547012	E	143.557052	E
143.501231	B	143.535012	G	143.547022	E	143.557062	E
143.501241	B	143.535022	G	143.547032	E	143.557072	E
143.501251	B	143.535062	G	143.551012	H	143.557082	E
143.501261	B	143.536012	F	143.551032	H	143.565022	G
143.501270	B	143.536022	F	143.551042	B	143.566002	F
143.501271	B	143.536032	F	143.551052	H	143.566012	F
143.502011	B	143.536042	F	143.551062	H	143.566022	F
143.502021	D	143.536052	F	143.551072	H	143.566032	F
143.502031	D	143.536062	F	143.551082	H	143.566042	F
143.502041	D	143.537012	E	143.551092	H	143.566052	F
143.504011	D	143.541012	H	143.551102	H	143.566062	F
143.505010	F	143.541022	B	143.551152	H	143.566072	F
143.505011	F	143.541032	B	143.551162	H	143.566082	F
143.506010	C	143.541042	H	143.551192	H	143.566092	F
143.521011	B	143.541052	H	143.554012	I	143.566102	F
143.521021	B	143.541062	H	143.554022	I	143.566112	F
143.521051	B	143.541072	B	143.554032	I	143.566122	F
143.521061	B	143.541082	B	143.554042	I	143.566132	F
143.521071	B	143.541102	B	143.554052	I	143.566142	F
143.521081	G	143.541112	H	143.554072	I	143.566152	F
143.521091	B	143.541122	H	143.554082	I	143.566162	F
143.521101	B	143.541132	H	143.555012	G	143.566172	F
143.521111	B	143.541142	H	143.555022	G	143.566182	F
143.521121	B	143.541152	H	143.555032	G	143.566192	F
143.521131	C	143.541162	B	143.555042	G	143.566202	F
143.524021	D	143.541172	H	143.555052	G	143.566212	G
143.524031	D	143.541182	H	143.556012	F	143.566222	F
143.524041	D	143.541192	H	143.556022	F	143.566232	F
143.524051	D	143.541202	H	143.556032	F	143.566242	F
143.524061	D	143.541212	B	143.556042	F	143.566252	F
143.524071	D	154.541222	H	143.556052	F	143.567012	E
143.524081	D	143.541232	B	143.556062	F	143.567022	E
143.525021	G	143.541252	B	143.556072	F	143.567032	E
143.526011	G	143.541262	B	143.556082	F	143.567042	E
143.526021	F	143.541282	H	143.556092	F	143.571002	H

Model number-to-column letter cross reference chart

HORIZONTAL CRAFTSMAN ENGINES (cont.)

Craftsman Engine Models	See Column	Craftsman Engine Models	See Column	Craftsman Engine Models	See Column	Craftsman Engine Models	See Column
143.571012	H	143.584052	I	143.594032	I		
143.571022	H	143.584062	I	143.594042	I		
143.571032	H	143.584072	I	143.594052	I		
143.571042	H	143.584082	I	143.594060	I		
143.571052	H	143.584102	I	143.594072	I		
143.571062	H	143.584112	I	143.594082	I		
143.571072	H	143.584122	I	143.594092	D		
143.571082	H	143.584132	I	143.594102	I		
143.571092	H	143.584142	I	143.595012	G		
143.571102	H	143.585012	G	143.595042	G		
143.571112	H	143.585032	G	143.596012	K		
143.571122	H	143.585042	G	143.596022	K		
143.571152	B	143.586012	F	143.596042	F		
143.571162	H	143.586022	F	143.596052	K		
143.571172	H	143.586032	F	143.596072	F		
143.574022	I	143.586042	F	143.596082	F		
143.574032	I	143.586052	F	143.596092	F		
143.574042	I	143.586062	F	143.596102	F		
143.574052	I	143.586072	F	143.596112	F		
143.574062	I	143.586082	F	143.596122	F		
143.574072	I	143.586112	K	143.597012	J		
143.574082	I	143.586122	K	143.597022	J		
143.574092	I	143.586132	K	143.597032	J		
143.574102	I	143.586142	K	143.601022	H		
143.575012	G	143.586152	F	143.601032	H		
143.575022	G	143.586162	K	143.604012	I		
143.575032	G	143.586172	F	143.604022	D		
143.575042	G	143.586182	F	143.604032	I		
143.576002	F	143.586192	F	143.604042	I		
143.576012	F	143.586202	F	143.604052	D		
143.576022	F	143.586212	F	143.604062	I		
143.576032	F	143.586222	F	143.605012	G		
143.576042	F	143.586242	F	143.605022	G		
143.576052	F	143.586252	K	143.606012	K		
143.576062	F	143.586262	F	143.606022	K		
143.576072	F	143.586272	F	143.606032	K		
143.576082	F	143.586282	F	143.606042	K		
143.576102	F	143.586232	F	143.606052	K		
143.576112	F	143.587012	J	143.607012	J		
143.576122	F	143.587022	J	143.607022	J		
143.576132	F	143.587032	J	143.607032	J		
143.576142	F	143.587042	J				
143.576172	F	143.591012	H				
143.576182	F	143.591022	H				
143.576192	F	143.591032	H				
143.576202	F	143.591042	H				
143.581002	H	143.591052	H				
143.581022	H	143.591062	H				
143.581032	H	143.591072	H				
143.581042	H	143.591082	H				
143.581052	H	143.591092	H				
143.581062	H	143.591102	H				
143.581072	H	143.591112	H				
143.581082	H	143.591122	H				
143.581092	H	143.591132	H				
143.581102	H	143.591142	H				
143.584012	I	143.594012	I				
143.584032	I	143.594022	I				

Model number-to-column letter cross reference chart

ESKA ENGINES

ESKA Model No.	See Column	ESKA Model No.	See Column	ESKA Model No.	See Column
ET25-10032	B	ET25K-2130	B	EW30G-2130	C
ET25A-1062	B	ET25K-2222	B	EW30G-2222	C
ET25A-1063	B	ET25K-2237	B	EW30G-2237	C
ET25A-1064	B	ET25K-2285	B		
ET25A-1065	B	ET25K-2310	B	EW30H-2130	C
ET25A-1066	B			EW30H-2222	C
ET25A-1280	B	ET25L-3169	B	EW30H-2237	
ET25A-1281	B	ET25L-3170	B	EW30H-2310	C
ET25A-1290	B				
ET25A-1303	B	ET25M-3169	B	EW30J-2130	C
		ET25M-3170	B	EW30J-2222	C
ET25B-1062	B			EW30J-2237	C
ET25B-1063	B	ET25N-3170	B	EW30J-2310	C
ET25B-1064	B				
ET25B-1065	B	ET27-10033A	B	EW30K-2130	C
ET25B-1066	B	ET27L-3171	B	EW30K-2222	C
ET25B-1280	B	ET27M-3171	B	EW30K-2237	C
ET25B-1281	B	ET27N-3171	B	EW30K-2310	C
ET25B-1290	B			EW30M-3400	C
ET25B-1303	B	ET30-30123A	H		
		ET30M-3099	C	HE25L-3383	B
ET25C-1062	B	ET30N-3099	C		
ET25C-1063	B			HT25N-2301	B
ET25C-1064	B	EW30A-1061	C	HT25P-2301	B
ET25C-1065	B	EW30A-1062	C		
ET25C-1066	B	EW30A-1063	C		
ET25C-1280	B	EW30A-1064	C		
ET25C-1281	B	EW30A-1065	C		
ET25C-1290	B	EW30A-1066	C		
ET25C-1303	B	EW30A-1280	C		
		EW30A-1281	C		
ET25D-1062	B	EW30A-1290	C		
ET25D-1063	B	EW30A-1303	C		
ET25D-1064	B				
FT25D-1065	B	EW30B-1061	C		
ET25D-1066	B	EW30B-1062	C		
ET25D-1280	B	EW30B-1063	C		
ET25D-1281	B	EW30B-1064	C		
ET25D-1290	B	EW30B-1065	C		
ET25D-1303	B	EW30B-1066	C		
		EW30B-1280	C		
ET25E-2130	B	EW30B-1281	C		
ET25E-2222	B	EW30B-1290	C		
ET25E-2237	B	EW30B-1303	C		
ET25G-2130	B	EW30D-1061	C		
ET25G-2222	B	EW30D-1062	C		
ET25G-2237	B	EW30D-1063	C		
		EW30D-1064	C		
ET25H-2130	B	EW30D-1065	C		
ET25H-2222	B	EW30D-1066	C		
ET25H-2237	B	EW30D-1280	C		
ET25H-2285	B				
ET25H-2310	B	EW30D-1281	C		
		EW30D-1290	C		
ET25J-2130	B	EW30D-1303	C		
ET25J-2222	B				
ET25J-2237	B	EW30E-2130	C		
ET25J-2285	B	EW30E-2222	C		
ET25J-2310	B	EW30E-2237	C		

Model number-to-column letter cross reference chart

REO ENGINES

REO Model No.	Ref. Letter	REO Model No.	Ref. Letter	REO Model No.	Ref. Letter	REO Model No.	Ref. Letter
M41	B	M66	B	MW8578I	H		
M42	B	M67	B	MW8578R	H		
M43	A	M67A	B	MW8579	H		
M44	C	M68	F	MW8582	H		
M45	E	M68A	F	MW8583	B		
M46	E	M69	B	MW8584	I		
M47	B	M70	C	MW8585	B		
M48	B	M71	B	MW8586	H		
M50	C	M72	C	MW8587	I		
M51	B	MW7704	B	MW8632	B		
M52	A	MW8278	B	MW8737	E		
M52A	B	MW8279	B	MW8741	H		
M53	B	MW8280	B	MW8742	H		
M54	C	MW8281	B	MW10597	I		
M54I	C	MW8282	B	MW10598	I		
M57	B	MW8283	B				
M58	B	MW8284	B				
M60	B	MW8286	G				
M61	C	MW8335	C				
M61A	C	MW8335	C				
M62	B	MW8395	B				
M63	B	MW8395	B				
M63A	B	MW8396	C				
M64	B	MW8396	C				
M65	B	MW8577	B				

TORO ENGINES

TORO Model No.	Ref. Letter	TORO Model No.	Ref. Letter	TORO Model No.	Ref. Letter	TORO Model No.	Ref. Letter
TE-1-2000	B	TE2-3011	C	TE3-20072C	H		
TE-1-2001	B	TE2-3014	C	TE3-20073C	H		
TE-1-2001A	B	TE2M-3015	C	TE3-20074C	H		
TE-1-2220	B	TE2M-3016	C	TE3-20109C	H		
TE-1-2221	B	TE2M-3416	C	TE3-20110D	H		
TE1-3010	B	TE2M-3417	C	TE3-30257D	H		
TE1-3011	B	TE2-30001	H				
TE1-3014	B	TE2-30001A	H				
TE1M-3015	B	TE2-30001C	H				
TE1M-3016	B	TE2-30071A	H				
TE1M-3416	B	TE2-30071C	H				
TE1M-3417	B						
TE1-10001	B	TE3-20043	H				
TE1-10002	B	TE3-20043C	H				
TE1-10002A	B	TE3-20043D	H				
TE1-10003	B	TE3-20044A	H				
TE1-10023A	B	TE3-20044C	H				
TE1-10028	B	TE3-20044D	H				
		TE3-20045A	H				
TE-2-2001	C	TE3-20045C	H				
TE-2-2001A	C	TE3-20045D	H				
TE-2-2220	C	TE3-20062C	H				
TE-2-2221	C	TE3-20064C	H				
TE2-3010	C	TE3-20069C	H				

Model number-to-column letter cross reference chart

FOUR CYCLE ENGINE
TABLE OF SPECIFICATIONS

Model	A	B	C	D	E	F	G	H	I
Displacement	6.207	7.35	7.61	8.9	11.04	13.53	12.176	7.75	9.06
Stroke	1-3/4"	1-3/4"	1-13/16"	1-13/16"	2-1/4"	2-1/2"	2-1/4"	1-27/32"	1-27/32"
Bore	2.125 / 2.127	2.3125 / 2.3135	2.3125 / 2.3135	2.5000 / 2.5010	2.5000 / 2.5010	2.6250 / 2.6260	2.6250 / 2.6260	2.3125 / 2.3135	2.5000 / 2.5010
Timing Dimension* Before Top Dead Center for Vertical Engines	V.050/.060	V.050/.060	V.050/.060	V.050/.060	V.080/.090	V.080/.090	V.080/.090	V.050/.060	V.050/.060
Timing Dimension* Before Top Dead Center for Horizontal Engines	H.060/.070	H.060/.070	H.060/.070	H.030/.040	H.090/.100	H.090/.100	H.090/.100	H.060/.070	H.030/.040
Point Setting	.020	.020	.020	.020	.020	.020	.020	.020	.020
Spark Plug Gap	.030	.030	.030	.030	.030	.030	.030	.030	.030
Valve Clearance	.010 Both	.010 Both	.010 Both	.010 Both	.010 Both	.010 Both	.010 Both	.010 Both	.010 Both
Valve Seat Angle	45°	30° Old 45° Cur	45°	45°	45°	45°	45°	45°	45°
Valve Spring Free Length	1"	1"	1"	1"	1-9/16"	1-9/16"	1-9/16"	1"	1"
Valve Spring Comp. Length	1/2"	1/2"	1/2"	1/2"	45/64"	45/64"	45/64"	1/2"	1/2"
Valve Guides Over-Size Dimensions	.2805 / .2815	.2805 / .2815	.2805 / .2815	.2805 / .2815	.3432 / .3442	.3432 / .3442	.3432 / .3442	.2805 / .2815	.2805 / .2815
Valve Seat Width	3/64"	3/64"	3/64"	3/64"	3/64"	3/64"	3/64"	3/64"	3/64"
Crankshaft End Play	.006 / .019	.006 / .019	.006 / .019	.006 / .019	.006 / .022	.006 / .022	.0015 / .0190	.006 / .019	.006 / .019

Engine specifications chart

FOUR CYCLE ENGINE
TABLE OF SPECIFICATIONS (Cont.)

J	K	L	M	N				Model
10.5	15.0	10.0	10.5	18.65				Displacement
1-15/16"	2-17/32"	1-27/32"	1-15/16"	2-17/32"				Stroke
2.6245 / 2.6255	2.750 / 2.751	2.625 / 2.626	2.625 / 2.626	3.062 / 3.063				Bore
V $\frac{.050}{.060}$	V $\frac{.080}{.090}$	.035	.035	V $\frac{.080}{.090}$				Timing Dimension* Before Top Dead Center for Vertical Engines
H $\frac{.060}{.070}$	H $\frac{.090}{.100}$							Timing Dimension* Before Top Dead Center for Horizontal Engines
.020	.020	.020	.020	.020				Point Setting
.030	.030	.030	.030	.030				Spark Plug Gap
.010 Both	.010 Both	.010 Both	.010 Both	.010 Both				Valve Clearance
45°	45°	46° $^{+1°}_{-0°}$	46° $^{+1°}_{-0°}$	46° $^{+1°}_{-0°}$				Valve Seat Angle
1"	1-9/16"	1.35±.03	1.35±.03	1-9/16"				Valve Spring Free Length
1/2"	45/64"	Solid Height .595	Solid Height .595	45/64"				Valve Spring Comp. Length
$\frac{.2805}{.2815}$	$\frac{.3432}{.3442}$	$\frac{.2807}{.2817}$	$\frac{.2807}{.2817}$	$\frac{.3432}{.3442}$				Valve Guides Over-Size Dimensions
3/64"	3/64"	$\frac{.035}{.045}$	$\frac{.035}{.045}$	$\frac{.042}{.052}$				Valve Seat Width
$\frac{.006}{.019}$	$\frac{.006}{.022}$	$\frac{.006}{.019}$	$\frac{.006}{.019}$	$\frac{.006}{.026}$				Crankshaft End Play

Engine specifications chart

FOUR CYCLE ENGINE
TABLE OF SPECIFICATIONS (Cont.)

Model	A	B	C	D	E	F	G	H	I
Crankpin Journal Diameter	.8610 .8615	.8610 .8615	.8610 .8615	.8610 .8615	1.0620 1.0625	1.0620 1.0625	1.0620 1.0625	.8610 .8615	.8610 .8615
Crankshaft Magneto Side Main Bearing Diameter	.8750 .8755	.8750 .8755	.8750 .8755	.8750 .8755	1.0005 1.0010	1.0005 1.0010	1.0005 1.0010	.8750 .8755	.8750 .8755
Crankshaft P.T.O. Side Main Bearing Diameter	.8750 .8755	.8750 .8755	.8750 .8755	.8750 .8755	1.0005 1.0010	1.0005 1.0010	1.0005 1.0010	.8750 .8755	.8750 .8755
Conn. Rod Dia. Crank Bearing	.8620 .8625	.8620 .8625	.8620 .8625	.8620 .8625	1.0630 1.0635	1.0630 1.0635	1.0630 1.0635	.8620 .8625	.8620 .8625
Piston Diameter	2.1210 2.1215	2.3097 2.3102	2.3097 2.3102	2.4925 2.4945	2.4925 2.4945	2.6200 2.6205	2.6200 2.6205	2.3097 2.3102	2.4925 2.4945
Piston Pin Diameter	.5629 .5631	.5629 .5631	.5629 .5631	.5629 .5631	.6248 .6250	.6248 .6250	.6248 .6250	.5629 .5631	.5629 .5631
Width Comp. Ring Groove	.0955 .0975	.0955 .0975	.0955 .0977	.0955 .0975	.0955 .0975	.0955 .0975	.0955 .0975	.0955 .0977	.0955 .0975
Width Oil Ring Groove	.125 .127	.125 .127	.125 .127	.125 .127	.1565 .1585	.1565 .1585	.1565 .1585	.125 .127	.125 .127
Side Clearance Ring Groove (Top) Comp. (Bot.) Oil	.002 .005	.002 .005	.002 .005	.0020 .0035	.0020 .0035	.0020 .0045	.0020 .0045	.002 .005	.0020 .0035
Ring End Gap	.007 .017	.007 .017	.007 .017	.007 .017	.007 .017	.007 .017	.007 .017	.007 .017	.007 .017
Top Piston Land Clearance	.0115 .0145	.0115 .0145	.0115 .0145	.015 .018	.015 .018	.017 .020	.017 .020	.0015 .0145	.015 .018
Piston Skirt Clearance	.0050 .0065	.0050 .0065	.0050 .0065	.0065 .0080	.0065 .0080	.0045 .0060	.0045 .0060	.0050 .0065	.0065 .0080
Camshaft Bearing Dia.	.4980 .4985	.4980 .4985	.4980 .4985	.9775 .9795	.6235 .6240	.6235 .6240	.6235 .6240	.4980 .4985	.497 .498
Cam Lobe Dia. Nose to Heel	.9775 .9795	.9775 .9795	.9775 .9795	.9775 .9795	1.258 1.262	1.258 1.262	1.258 1.262	.9775 .9795	.9775 .9795
Dia. Crankshaft Mag. Journal	.8735 .8740	.8735 .8740	.8735 .8740	.8735 .8740	.9985 .9990	.9985 .9990	.9985 .9990	.8735 .8740	.8735 .8740
Dia. Crankshaft P.T.O. Journal	.8735 .8740	.8735 .8740	.8735 .8740	.8735 .8740	.9985 .9990	.9985 .9990	.9985 .9990	.8735 .8740	.8735 .8740

Engine specifications chart

FOUR CYCLE ENGINE
TABLE OF SPECIFICATIONS (Cont.)

J	K	L	M	N			Model
.9995 / 1.0000	1.1865 / 1.1870	.8610 / .8615	.9995 / 1.0000	1.1865 / 1.1870			Crankpin Journal Diameter
1.0005 / 1.0010	1.0005 / 1.0010	.8755 / .8760	1.0005 / 1.0010	1.0005 / 1.0010			Crankshaft Magneto Side Main Bearing Diameter
1.0005 / 1.0010	1.0005 / 1.0010	1.2010 / 1.2020	1.2010 / 1.2020	1.1890 / 1.1895			Crankshaft P.T.O. Side Main Bearing Diameter
1.0005 / 1.0010	1.1880 / 1.1885	.8620 / .8625	1.0005 / 1.0010	1.1880 / 1.1885			Conn. Rod Dia. Crank Bearing
2.6185 / 2.6205	2.7427 / 2.7442	2.604 / 2.608	2.604 / 2.608	3.0547 / 3.0562			Piston Diameter
.5631 / .5635	.6250 / .6254	.5631 / .5635	.5631 / .5635	.6250 / .6254			Piston Pin Diameter
.0925 / .0935	.0795 / .0805	.0955 / .0975	.0955 / .0975	.0955 / .0975			Width Comp. Ring Groove
.1545 / .1555	.1880 / .1890	.1565 / .1585	.1565 / .1585	.188 / .190			Width Oil Ring Groove
.0020 / .0040 .0010 / .0040	.0020/.0035 .0015/.0035 .0010/.0030	.0020 / .0050 .0015 / .0040	.0020 / .0050 .0015 / .0040	.003 / .004 .002 / .003			Side Clearance Ring Groove — (Top) Comp. (Bot.) Oil
.016 / .020	.010 / .020	.010 / .020	.010 / .020	.010 / .020			Ring End Gap
.0165 / .0215	.023 / .028	.017 / .022	.017 / .022	.031 / .034			Top Piston Land Clearance
.0050 / .0055	.0045 / .0060	.0045 / .0075 .0045 / .0060	.0045 / .0075 .0045 / .0060	.005 / .007			Piston Skirt Clearance
.4975 / .4980	.6235 / .6240	.505 / .513	.505 / .513	.6235 / .6240			Camshaft Bearing Dia.
Ex. .9775 / .9795 In. 1.2939 / 1.2959	1.267 / 1.263	Ex. .9775 / .9795 In. .9775 / .9795	.9795 / .9775 1.2959 / 1.2939				Cam Lobe Dia. Nose to Heel
.9985 / .9990	.9985 / .9990	.8735 / .8740	.9985 / .9990	.9985 / .9990			Dia. Crankshaft Mag. Journal
.9985 / .9990	.9985 / .9990	.8735 / .8740	.9985 / .9990	1.1870 / 1.1875			Dia. Crankshaft P.T.O. Journal

Engine specifications chart

FOUR CYCLE TORQUE SPECIFICATIONS

		INCH POUNDS	FT. POUNDS
Cylinder Head Bolts		140 - 200	12 - 16
Connecting Rod Lock Nuts	1.5-3.5 H.P.	65 - 75	5.5 - 6
	4 - 5 H.P. Small Frame	80 - 95	6.6 - 7.9
	4 - 6 H.P. Medium Frame	86 - 110	7 - 9
	7 - 8 H.P. Medium Frame	106 - 130	8.8 - 10.8
Cylinder Cover or Flange to Cylinder		65 - 110	5.5 - 9
Flywheel Nut		360 - 400	30 - 33
Spark Plug		180 - 240	15 - 20
Magneto Stator to Cylinder		60 - 84	5 - 7
Starter to Blower Housing		40 - 60	3.5 - 5
Blower Housing to Crankcase		48 - 72	4 - 6
Breather Cover		20 - 26	1.7 - 2.1
Intake Pipe to Cylinder		72 - 96	6 - 8
Carburetor to Intake Pipe		48 - 72	4 - 6
Air Cleaner to Carburetor		15 - 25	1 - 1.5
Fuel Tank Mounting		110 - 130	9 - 10
Muffler Bolts to Cylinder	1 - 5 H.P. Small Frame	30 - 45	2.5 - 3.5
	4 - 8 H.P. Medium Frame	90 - 150	8 - 12
6:1 Gear Reduction Housing to Cylinder		100 - 144	8.5 - 12
Gear Reduction Cover to Housing		65 - 110	5 - 9
Drain Plug (engine)		65 - 100	5.5 - 8.4
Locked Ball Bearing Retainer Nut		15 - 22	1.5
Craftsman Exclusive Fuel System to Cylinder		72 - 96	6 - 8

Torque specifications chart

2-Stroke Engines

ENGINE IDENTIFICATION

The identification tags may be located at a variety of places on the engine. The type number is the most important number since it must be included with any correspondence about a particular engine.

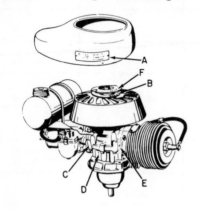

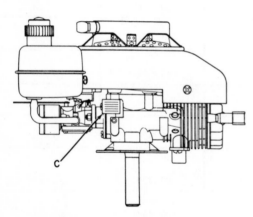

A. Nameplate on Air Shroud
B. Model & Type Number Plate
C. Metal Tag on Crankcase
D. Stamped on Crankcase
E. Stamped on Cylinder Flange
F. Stamped on Starter Pulley

Location of identification numbers on 2 cycle engines

Early engines listed the type number as a suffix of the serial number. For example on number 123456789 P 234, 234 is the type number. In the number 123456789 H 104-02B; 104-02B is the type number. In either case the type number is important.

If you use short block to repair the en-gine, be sure that you transfer the serial number and type number tag to the new short block.

On the newer engines, reference is sometimes made to the model number. The model number tells the number of cylinders, the design (vertical or horizontal) and the cubic inch displacement.

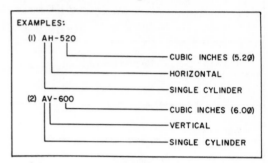

Model number interpretation

FUELS AND LUBRICANTS

Tecumseh 2 cycle engines are mist-lubricated by oil mixed with the gasoline. For the best performance, use regular grade, leaded fuel, with 2 cycle or outboard oil rated SAE 30 or SAE 40. The terms 2 cycle or outboard are used by various manufacturers to designate oil they have designed for use in 2 cycle engines. multiple weight oil such as all season 10W–30, are not recommended.

If you have to mix the gas and oil when the temperature is below 35° F, heat up the oil first, then mix it with the gas. Oil will not mix with gas when the temperature is approaching freezing. However, if you use oil that has been warmed first it will not be affected by low temperatures. Double check the manufacturer's fuel mixture recommendations in the owner's manual.

AIR CLEANERS

All polyurethane air filters and aluminum foil filters should be cleaned in warm soapy water, dried, soaked in oil, and then installed on the engine after allowing excess oil to drop off.

Felt and fiber elements should be cleaned by blowing compressed air through the filter in the opposite direction of normal air flow. Clean the covers and mounting brackets with a damp cloth.

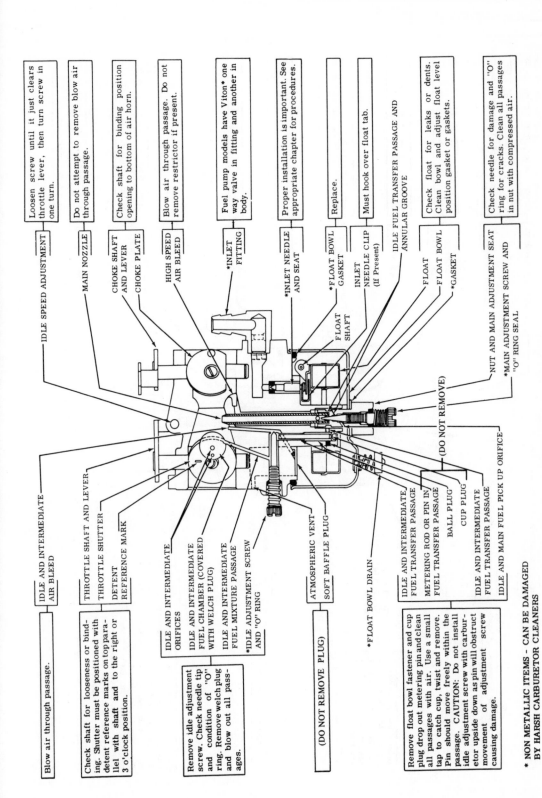

Loosen screw until it just clears throttle lever, then turn screw in one turn.

Do not attempt to remove air blow through passage.

Check shaft for binding position opening to bottom of air horn.

Blow air through passage. Do not remove restrictor if present.

Fuel pump models have Viton* one way valve in fitting and another in body.

Proper installation is important. See appropriate chapter for procedures.

Replace.

Must hook over float tab.

Check float for leaks or dents. Clean bowl and adjust float level position gasket or gaskets.

Check needle for damage and "O" ring for cracks. Clean all passages in nut with compressed air.

IDLE SPEED ADJUSTMENT

MAIN NOZZLE

CHOKE SHAFT AND LEVER

CHOKE PLATE

HIGH SPEED AIR BLEED

*INLET FITTING

*INLET NEEDLE AND SEAT

*FLOAT BOWL GASKET

INLET NEEDLE CLIP (If Present)

IDLE FUEL TRANSFER PASSAGE AND ANNULAR GROOVE

FLOAT SHAFT

*FLOAT

FLOAT BOWL

*GASKET

NUT AND MAIN ADJUSTMENT SEAT

*MAIN ADJUSTMENT SCREW AND "O" RING SEAL

(DO NOT REMOVE)

Blow air through passage.

Check shaft for looseness or binding. Shutter must be positioned with detent reference marks on top parallel with shaft and to the right or 3 o'clock position.

Remove idle adjustment screw. Check needle tip and condition of "O" ring. Remove welch plug and blow out all passages.

IDLE AND INTERMEDIATE AIR BLEED

THROTTLE SHAFT AND LEVER

THROTTLE SHUTTER

DETENT REFERENCE MARK

IDLE AND INTERMEDIATE ORIFICES

IDLE AND INTERMEDIATE FUEL CHAMBER (COVERED WITH WELCH PLUG)

IDLE AND INTERMEDIATE FUEL MIXTURE PASSAGE

*IDLE ADJUSTMENT SCREW AND "O" RING

ATMOSPHERIC VENT

SOFT BAFFLE PLUG

(DO NOT REMOVE PLUG)

*FLOAT BOWL DRAIN

IDLE AND INTERMEDIATE FUEL TRANSFER PASSAGE

METERING ROD OR PIN IN FUEL TRANSFER PASSAGE

BALL PLUG

CUP PLUG

IDLE AND INTERMEDIATE FUEL TRANSFER PASSAGE

IDLE AND MAIN FUEL PICK UP ORIFICE

Remove float bowl fastener and cup plug drop out metering pin and clean all passages with air. Use a small tap to catch cup, twist and remove. Pin should move freely within the passage. CAUTION: Do not install idle adjustment screw with carburetor upside down as pin will obstruct movement of adjustment screw causing damage.

* NON METALLIC ITEMS - CAN BE DAMAGED BY HARSH CARBURETOR CLEANERS

Service procedures for float feed carburetors

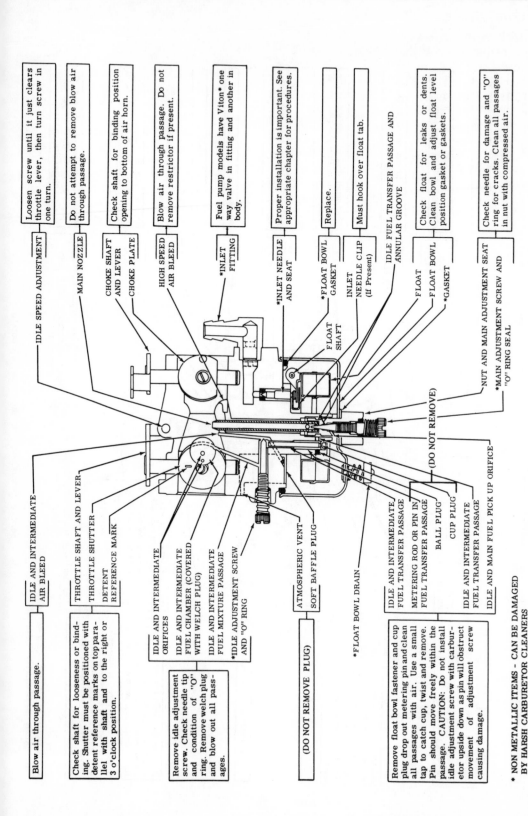

IDLE SPEED ADJUSTMENT — Loosen screw until it just clears throttle lever, then turn screw in one turn.

MAIN NOZZLE — Do not attempt to remove blow air through passage.

CHOKE SHAFT AND LEVER — Check shaft for binding position opening to bottom of air horn.

CHOKE PLATE

HIGH SPEED AIR BLEED — Blow air through passage. Do not remove restrictor if present.

INLET FITTING — Fuel pump models have Viton one way valve in fitting and another in body.

*INLET NEEDLE AND SEAT — Proper installation is important. See appropriate chapter for procedures.

*FLOAT BOWL GASKET — Replace.

INLET NEEDLE CLIP (If Present) — Must hook over float tab.

IDLE FUEL TRANSFER PASSAGE AND ANNULAR GROOVE

FLOAT — Check float for leaks or dents. Clean bowl and adjust float level position gasket or gaskets.

FLOAT BOWL

*GASKET

NUT AND MAIN ADJUSTMENT SEAT

*MAIN ADJUSTMENT SCREW AND "O" RING SEAL — Check needle for damage and "O" ring for cracks. Clean all passages in nut with compressed air.

IDLE AND INTERMEDIATE AIR BLEED — Blow air through passage.

THROTTLE SHAFT AND LEVER — Check shaft for looseness or binding. Shutter must be positioned with detent reference marks on top parallel with shaft and to the right or 3 o'clock position.

THROTTLE SHUTTER

DETENT REFERENCE MARK

IDLE AND INTERMEDIATE ORIFICES

IDLE AND INTERMEDIATE FUEL CHAMBER (COVERED WITH WELCH PLUG) — Remove idle adjustment screw. Check needle tip and condition of "O" ring. Remove welch plug and blow out all passages.

IDLE AND INTERMEDIATE FUEL MIXTURE PASSAGE

*IDLE ADJUSTMENT SCREW AND "O" RING

FLOAT SHAFT

ATMOSPHERIC VENT

SOFT BAFFLE PLUG — (DO NOT REMOVE PLUG)

*FLOAT BOWL DRAIN

IDLE AND INTERMEDIATE FUEL TRANSFER PASSAGE

METERING ROD OR PIN IN FUEL TRANSFER PASSAGE — (DO NOT REMOVE)

BALL PLUG

CUP PLUG — Remove float bowl fastener and cup plug drop out metering pin and clean all passages with air. Use a small tap to catch cup, twist and remove. Pin should move freely within the passage. CAUTION: Do not install idle adjustment screw with carburetor upside down as pin will obstruct movement of adjustment screw causing damage.

IDLE AND INTERMEDIATE FUEL TRANSFER PASSAGE

IDLE AND MAIN FUEL PICK UP ORIFICE

* NON METALLIC ITEMS - CAN BE DAMAGED BY HARSH CARBURETOR CLEANERS

Service procedures for diaphragm carburetors

CARBURETORS

All 2 cycle engines use either float feed type carburetors or diaphragm type carburetors. All carburetors used on 2 cycle engines are similar in design and operation to those used on the 4 cycle engines. Use the same procedures for servicing the 2 cycle carburetors as for the 4 cycle carburetors.

NOTE: *The float adjustment for Tillotson float carburetors is* $1^{13}/_{32} \pm \frac{1}{64}$ *in. with the needle valve seated.*

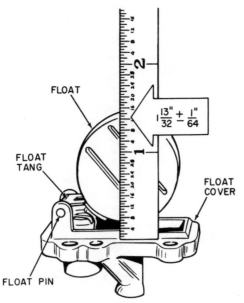

Float adjustment for Tillotson MT-50A and MT-51A carburetors

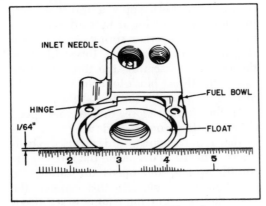

Float adjustment for Tillotson MD carburetor

GOVERNORS

Idle Governor

The idle governor controls the throttle shutter movement at idle speeds only. It reduces fuel "puddling" within the crankcase, thus allowing the engine to idle smoothly for prolonged periods of time.

Air rushing through the air horn strikes the offset throttle shutter and closes the throttle. The spring tension in the idle governor has a tendency to hold the throttle slightly open. As the closed throttle reduces the air velocity, the spring tension on the throttle shaft overcomes the reduced pressure of the air and reopens the throttle.

To service the idle governor, remove the throttle lever and plate and check and replace worn parts.

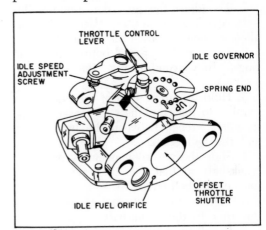

Idle governor

The throttle valve is closed by air striking it.

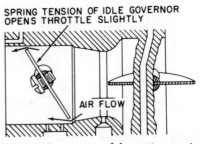

The throttle valve is opened by spring tension.

Adjusting Power Take-Off End Mechanical Governors

The governor assembly consists of two rings, weighted linkage, and the governor spring. The lower ring is secured to the crankshaft so the governor assembly rotates with the crankshaft. As rotational speed increases, the weighted linkages are thrown outward by centrifugal force. The top ring must move toward the bottom ring. The governor bell crank which contacts the top ring follows it, permitting the throttle to close and slow the engine speed. As centrifugal force decreases, the governor spring moves the top ring upward, overcomes the throttle spring, and opens the throttle while maintaining a constant operating engine speed.

1. To adjust the governor, remove the outboard bearing housing. Use a ³⁄₃₂ in. allen wrench to loosen the set screw.

2. Squeeze the top and bottom governor rings, fully compressing the governor spring.

3. Hold the upper arm of the bell crank parallel to the crankshaft and insert a ³⁄₃₂ in. allen wrench between the upper ring and the bell crank.

4. Slide the governor assembly onto the crankshaft so that the allen wrench just touches the bell crank.

5. Tighten the set screw to secure the governor to the crankshaft.

6. Install the bearing adapter, mount the engine, and check the engine speed with a tachometer. It should be about 3200 to 3400 rpm.

NOTE: *Never attempt to adjust the governor by bending the bellcrank or the link.*

7. If the speed is not correct, readjust the governor by moving the assembly toward the crankcase to increase speed, and away from the crankcase to decrease speed.

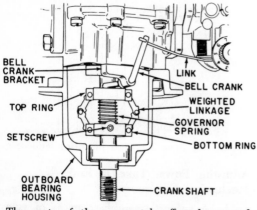

The parts of the power take-off end mounted governor (cutaway view)

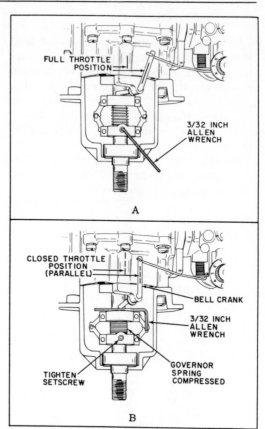

Adjusting the PTO end mounted governor

Mechanical Flywheel Type Governor Adjustment

As engine speed increases, the links are thrown outward, compressing the link springs. The links apply a thrust against the slide ring, moving it upward and compressing the governor spring. As the slide ring moves away from the thrust block of the bellcrank assembly, the throttle spring causes the thrust block to maintain engagement and close the throttle slightly.

As the throttle closes and engine speed decreases, force on the slide ring decreases so that it moves downward, pivots the bellcrank outward to overcome the force of the throttle spring, and opens the throttle to speed up the engine. In this manner, the operating speed of the engine is stabilized to the adjusted governor setting.

1. To adjust the governor, loosen the bracket screw and slide the governor bellcrank assembly toward or away from the flywheel. Move the bellcrank toward the flywheel to increase speed and away from the flywheel to decrease speed.

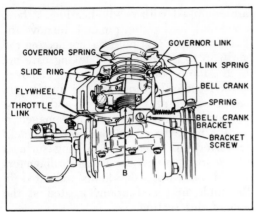

Flywheel mounted governor assembly (cutaway view)

2. Tighten the screw to secure the bracket.

3. Make minor speed adjustments by bending the throttle link at the bend in the center of the link.

NOTE: *Do not lubricate the governor assembly or the governor bellcrank assembly of flywheel mounted governors.*

Adjusting 2 Cycle Air Vane Governors

1. Loosen the self-locking nut that holds the governor spring bracket to the engine crankcase.

2. Adjust the spring bracket to increase or decrease the governor spring tension. Increasing spring tension increases speed and decreasing spring tension decreases speed.

3. After adjusting, the spring bracket

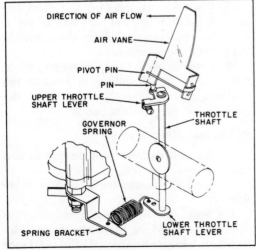

Diagram of an air vane governor

should not be closer than $\frac{1}{16}$ in. to the crankcase.

4. Tighten the self-locking nut.

IGNITION SYSTEM

The ignition systems on 2 cycle engines are similar enough to those systems on 4 cycle engines that they can be serviced in the same manner.

ENGINE MECHANICAL

Disassembly

SPLIT CRANKCASE ENGINES

1. Remove the shroud and fuel tank if so equipped.

2. Remove the flywheel and the ignition stator.

3. Remove the carburetor and governor linkage. Carefully note the position of the carburetor wire links and springs for re-installation.

4. Lift off the reed plate and gasket if present and inspect them. They should not bend away from the sealing surface plate more than 0.010 in.

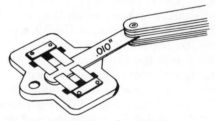

Checking the reed valve clearance

5. Remove the spark plug and inspect it.

6. Remove the muffler. Be sure that the muffler and the exhaust ports are not clogged with carbon. Clean them if necessary.

7. Remove the transfer port cover and check for a good seal.

8. Remove the cylinder head, if so equipped.

NOTE: *Some models utilize a locking compound on the cylinder head screws. Removing the screws on such engines can be difficult. This is especially true with screws having a slotted head for a straight screwdriver blade. The screws can be removed if heat is applied to the head of the screw with an electric soldering iron.*

9. On engines having a governor mounted on the power take-off end of the crankshaft, remove the screws that hold the outboard bearing housing to the crankcase. Clean the PTO end of the crankshaft and remove the outboard bearing housing and bearing. Loosen the set screw that holds the governor assembly to the crankshaft, slide the entire governor assembly from the crankshaft. Remove the screw that holds the governor bellcrank bracket to the crankcase. Remove the governor bellcrank and bracket.

10. Make match marks on the cylinder and crankcase. Remove the four nuts and lockwashers that hold the cylinder to the crankcase.

11. Remove the cylinder by pulling it straight out from the crankcase.

12. To separate the two crankcase halves, remove all of the screws that hold the crankcase halves together.

13. With the crankcase in a vertical position, grasp the top half of the crankcase and hold it firmly. Strike the top end of the crankcase with a rawhide mallet, while holding the assembly over a bench to prevent damage to parts when they fall. The top half of the crankcase should separate from the remaining assembly.

14. Invert the assembly and repeat the procedure to remove the other casting half from the crankcase on ball bearing units.

15. Each time the crankshaft is removed from the crankcase, seals at the end of the crankcase should be replaced. To replace the seals, use a screwdriver or an ice pick to remove the seal retainers and remove and discard the old seals. Install the seals in the bores of the crankcase halves. The seals must be inserted into the bearing well with the channel groove toward the internal side of the crankcase. Retain the seal with the retainer. Seat the retainer spring into the spring groove.

Connecting Rod Service

1. For engines using solid bronze or aluminum connecting rods, remove the two self-locking capscrews which hold the connecting rod to the crankshaft and remove the rod cap. Note the match marks on the connecting rod and cap. These marks must be reinstalled in the same position to the crankshaft.

2. Engines using steel connecting rods are equipped with needle bearings at both crankshaft and piston pin end. Remove the two set screws that hold the connecting rod and cap to the crankshaft, taking care not to lose the needle bearings during removal.

3. Needle bearings at the piston pin end of steel rods are caged and can be pressed out as an assembly if damaged.

4. Check the connecting rod for cracks or distortion. Check the bearing surfaces for scoring or wear. Bearing diameters should be within the limits indicated in the table of specifications located at the end of this section.

5. There are two basic arrangements of needles supplied with the connecting rod crankshaft bearing: split rows of needles and a single row of needles. Service needles are supplied with a beeswax coating. The beeswax holds the needles in position.

6. To install the needle bearings, first make sure that the crankshaft bearing journal and the connecting rod are free from oil and dirt.

7. Place the needle bearings with the beeswax onto a cool metallic surface to stiffen the beeswax. Body temperature will melt the wax, so avoid handling.

8. Remove the paper backing on the bearings and wrap the needles around the crankshaft journal. The beeswax will hold the needles onto the journal. Position the needles uniformly onto the crankpin.

NOTE: *When installing the split row of needles, wrap each row of needles around the journal and try to seal them together with gentle but firm pressure to keep the bearings from unwinding.*

9. Place the connecting rod onto the journal, position the rod cap, and secure it with the capscrews. Tighten the screw to the proper specifications.

10. Force solvent (lacquer thinner) into the needles just installed to remove the beeswax, then force 30W oil into the needles for proper lubrication.

Piston and Rings Service

1. Clean all carbon from the piston and ring grooves.

2. Check the piston for scoring or other damage.

3. Check the fit of the piston in the cylinder bore. Move the piston from side-to-side to check clearance. If the clearance is not greater than 0.003 in. and the cylin-

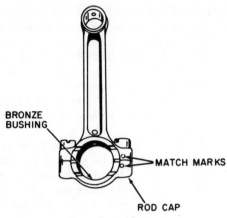

Connecting rod match marks

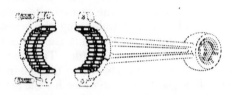

A. SPLIT ROWS OF NEEDLE BEARINGS

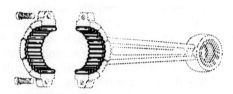

B. SINGLE ROW OF NEEDLE BEARINGS

Needle bearing arrangement. Double rows of bearings are placed with the tapered edges facing out.

der is not scored or damaged, then the piston need not be replaced.

4. Check the piston ring side clearance to make sure it is within the limits recommended.

5. Check the piston rings for wear by inserting them into the cylinder about ½ in. from the top of the cylinder. Check at various places to make sure that the gap between the ends of the ring does not exceed the dimensions recommended in the specifications table at the end of this section. Bore wear can be checked in the same way, except that a new ring is used to measure the end gap.

6. If replacement rings have a bevelled or chamfered edge, install them with the bevel up toward the top of the piston. Not all engines use bevelled rings. The two rings installed on the piston are identical.

7. When installed, the offset piston used on the AV600 and the AV520 engines must have the "V" stamped in the piston head (some have hash marks) facing toward the right as the engine is viewed from the top or piston side of the engine.

NOTE: *Some AV520 and AV600 engines do not have offset pistons. Only offset pistons will have the "V" or the hash marks on the piston head. Domed pistons must be installed so that the slope of the piston is toward the exhaust port.*

Crankshaft Service

1. Use a micrometer to check the bearing journals for out-of-roundness. The main bearing journals should not be more than 0.0005 in. out-of-round. Connecting rod journals should not be more that 0.001 in. out-of-round. Replace a crankshaft that is not within these limits.

NOTE: *Do not attempt to regrind the crankshaft since undersize parts are not available.*

2. Check the tapered portion of the crankshaft (magneto end), keyways, and threads. Damaged threads may be restored with a thread die. If the taper of the shaft is rusty, it indicates that the engine has been operating with a loose flywheel. Clean the rust off the taper and check for wear. If the taper or keyway is worn, replace the crankshaft.

3. Check all of the bearing journal diameters. They should be within the limits indicated in the specifications table at the end of this section.

4. Check the crankshaft for bends by placing it between two pivot points. Position the dial indicator feeler on the crank-

Checking the crankshaft for out-of-roundness

shaft bearing surface and rotate the shaft. The crankshaft should not be more than 0.002 to 0.004 in. out-of-round.

Bearing Service

1. Do not remove the bearings unless they are worn or noisy. Check the operation of the bearings by rotating the bearing cones with your fingers to check for roughness, binding, or any other signs of unsatisfactory operation. If the bearings do not operate smoothly, remove them.

2. To remove the bearings from the crankcase, the crankcase must be heated. Use a hot plate to heat the crankcase to no more than 400° F. Place a ⅛ in. steel plate over the hot plate to prevent overheating. At this temperature, the bearings should drop out with a little tapping of the crankcase.

3. The replacement bearing is left at room temperature and dropped into the heated crankcase. Make sure that the new bearing is seated to the maximum depth of the cavity.

NOTE: *Do not use an open flame to heat the crankcase halves and do not heat the crankcase halves to more than 400° F. Uneven heating with an open flame or excessive temperature will distort the case.*

4. The needle bearings will fall out of the bearing cage with very little urging. Needles can be reinstalled easily by using a small amount of all-purpose grease to hold the bearings in place.

5. Cage bearings are removed and replaced in the same manner as ball bearings.

6. Sleeve bearings cannot be replaced. Both crankcase halves must be discarded if a bearing is worn excessively.

Assembly

1. The gasket surface where the crankcase halves join must be thoroughly clean before reassembly. Do not buff or use a file or any other abrasive that might damage the mating surfaces.

NOTE: *Crankcase halves are matched. If one needs to be replaced, then both must be replaced.*

2. Place the PTO half of the crankcase onto the PTO end of the crankshaft. Use seal protectors where necessary.

3. Apply a thin coating of sealing compound to the contact surface of one of the crankcase halves.

4. Position one crankcase half on the other. The fit should be such that some pressure is required to bring the two halves together. If this is not the case, either the crankcase halves and/or the crankshaft must be replaced.

5. Secure the halves with the screws provided, tightening the screws alternately and evenly. Before tightening the screws, check the union of the crankcase halves on the cylinder mounting side. The halves should be flat and smooth at the union to provide a good mounting face for the cylinder. If necessary, realign the halves before tightening the screws.

6. The sleeve tool should be placed into the crankcase bore from the direction opposite the crankshaft. Insert the tapered end of the crankshaft through the half of the crankcase to which the magneto stator is mounted. Remove the seal tools after installing the crankcase halves.

7. Stagger the ring ends on the piston and check for the correct positioning on domed piston models.

8. Place the cylinder gasket on the crankcase end.

9. Place the piston into the cylinder using the chamfer provided on the bottom edge of the cylinder to compress the rings.

10. Secure the cylinder to the crankcase assembly.

STARTERS

Two cycle Tecumseh engines use the same recoil starter as the four cycle engines. Refer to the four cycle engine section for starter procedures.

Type No.	Column Letter		Type No.	Column Letter		Type No.	Column Letter	
1 thru 44	A	SB	401 thru 402	L	SB	618-17	F	SB
46 thru 68	B	SB	403	M	SB	618-18	E	SB
69 thru 77	D	SB	404 thru 406B	L	SB	619 thru 619-03	E	SB
78 thru 80	C	SB	407 thru 407C	M	SB	621 thru 621-11	E	SB
81 thru 83	D	SB	408 thru 410B	L	SB	621 thru 621-12A	F	SB
84 thru 85	C	SB	411 thru 411B	M	SB	621-13 thru 621-15	E	SB
86 thru 87	B	SB	412 thru 423	L	SB	622 thru 622-07	E	SB
89	C	SB	424 thru 427	M	SB	623 and 623A	E	SB
91	D	SB				623-01	F	SB
93	D	SB				623-02 thru 623-33E	E	SB
94 thru 98	H	SB				623-34	F	SB
99	C	SB	501 thru 509A	E	SB	623-35 and 623-36	E	SB
						624 thru 630-09	J	SB
						632-02A	K	SB
						632-04 and 632-05A	K	SB
201 thru 208	A	SB	601 thru 601-01	F	SB	633 thru 634-09	J	SB
209 thru 244	B	SB	602 thru 602-02B	F	SB	635A and B	K	SB
245	D	SB	603A thru 603-23	E	SB	635-03B	K	SB
246 thru 248	B	SB	604 thru 604-25	E	SB	635-04B	K	SB
249 thru 251	D	SB	605 thru 605-18	E	SB	635-06A thru 635-10	K	SB
252	B	SB	606 thru 606-05B	E	SB	636 thru 636-11	J	SB
253 thru 255	D	SB	606-06 thru 606-07	F	SB	637 thru 637-16	J	SB
256 thru 261	B	SB	606-08 thru 606-13	E	SB	638 thru 638-100	F	UB
262 thru 265A	D	SB	608 thru 608-C	E	SB	639	M	UB
266 thru 267	B	SB	610-A thru 610-14	E	SB	640	O	UB
268 thru 272	D	SB	610-15 thru 610-16	F	SB	641	K	UB
273 thru 275	B	SB	610-19 thru 610-20	E	SB	642	I	UB
276	D	SB	611	F	SB	643	J	UB
277 thru 279	B	SB	614-01 thru 614-04	E	SB	650	N	UB
280 thru 281	D	SB	614-05 thru 614-06A	F	SB	670	H	UB
282	B	SB	615-01 thru 615-01A	E	SB			
283 thru 284	D	SB	615-04	F	SB			
285 thru 286	B	SB	615-05 thru 615-10A	E	SB			
287	D	SB	615-11 thru 615-18	F	SB			
288 thru 291B	B	SB	615-19 thru 615-27A	E	SB			
292 thru 294	D	SB	615-28	F	SB			
295 thru 296	B	SB	615-29 thru 615-39	E	SB			
297	F	SB	616-01 thru 616-11	E	SB			
298 thru 299	B	SB	616-12 and 616-14	F	SB			
			616-16 thru 616-40	E	SB			
			617A thru 617-01	E	SB			
			617-02 thru 617-03A	F	SB			
301 thru 375			617-04 thru 617-04A	E	SB			
These are twin			617-05 thru 617-05A	F	SB			
cylinder units -			617-06	E	SB			
out of production			618 thru 618-16	E	SB			

Type number-to-letter cross reference chart for use with the following specifications charts

Type No.	Column Letter		Type No.	Column Letter		Type No.	Column Letter	
701 thru 701-1C	G	SB	1055 thru 1056B	G	SB	1238 thru 1238C	H	SB
701-2 thru 701-2D	H	SB	1057	H	SB	1239 thru 1243	J	SB
701-3 thru 701-3C	G	SB	1058 thru 1058A	G	SB	1244 thru 1245A	K	SB
701-4 thru 701-4C	H	SB	1059	D	SB	1246 thru 1247	J	SB
701-5 thru 701-5C	G	SB	1060 thru 1060C	H	SB	1248	H	SB
701-6 thru 701-6C	H	SB	1061 and 1062	C	SB	1249 thru 1251A	J	SB
701-7 thru 701-7C	G	SB	1063A thru 1063C	H	SB	1252 thru 1254A	K	SB
701-8 thru 701-11	H	SB	1064	G	SB	1255	J	SB
701-12 thru 701-13A	G	SB	1065 thru 1067C	D	SB	1256 thru 1262B	K	SB
701-14 thru 701-17	H	SB	1068 and 1068A	G	SB	1263	J	SB
701-18 thru 701-19	G	SB	1069	H	SB	1264 thru 1265E	K	SB
701-19A	H	SB	1070 thru 1070B	C	SB	1266 thru 1267	G	SB
701-20 thru 701-22A	G	SB	1071 thru 1075D	H	SB	1269 thru 1270D	K	SB
701-24	H	SB	1076 thru 1084	H	SB	1271 thru 1271B	G	SB
			1085 and 1085A	G	SB	1272 thru 1275	K	SB
			1086 thru 1140A	H	SB	1276	H	SB
			1141 thru 1142	G	SB	1277 thru 1279D	K	SB
1001	D	SB	1143 thru 1145	H	SB	1280	J	SB
1002 thru 1002D	G	SB	1146 thru 1149A	I	SB	1283 thru 1284D	K	SB
1003	C	SB	1149A thru 1153B	H	SB	1286 thru 1286A	H	SB
1004 thru 1004A	H	SB	1158 thru 1159	G	SB	1287	K	SB
1005 thru 1005A	C	SB	1160 thru 1161A	I	SB	1288	J	SB
1006	H	SB	1162 and 1162A	G	SB	1289 thru 1289A	G	SB
1007 thru 1008B	C	SB	1163 thru 1163A	H	SB	1290 thru 1293A	K	SB
1009 thru 1010B	G	SB	1165 thru 1168A	G	SB	1294 thru 1295	J	SB
1011 thru 1019E	H	SB	1169 thru 1174	H	SB	1296 thru 1298	K	SB
1020	G	SB	1175 and 1175A	I	SB			
1021	C	SB	1176 thru 1177A	H	SB			
1022 thru 1022C	D	SB	1179 thru 1179A	I	SB			
1023 thru 1026A	H	SB	1180 thru 1181	G	SB	1300 thru 1303	K	SB
1027 thru 1027B	C	SB	1182 and 1182A	I	SB	1304 thru 1307	J	SB
1028 thru 1030C	H	SB	1183 thru 1185	G	SB	1308 thru 1316A	K	SB
1031B thru 1031C	G	SB	1186A thru 1186C	J	SB	1317 thru 1317B	J	SB
1032	C	SB	1187 thru 1192	G	SB	1318 thru 1320B	K	SB
1033 thru 1033F	G	SB	1192A thru 1196B	J	SB	1321 thru 1322	J	SB
1034 thru 1034G	H	SB	1197 thru 1197A	G	SB	1323	K	SB
1035 thru 1036	C	SB	1198 and 1198A	H	SB	1325	G	SB
1037 thru 1039C	D	SB	1199 and 1199A	J	SB	1326 thru 1326F	K	SB
1040 thru 1041A	D	SB	1206 thru 1208B	J	SB	1327 thru 1327B	H	SB
1042 thru 1042C	G	SB	1210 thru 1215C	K	SB	1328B and 1328C	K	SB
1042D thru 1042E	H	SB	1216 thru 1216A	H	SB	1329	J	SB
1042F	G	SB	1217 thru 1220A	J	SB	1330	H	SB
1042G and 1042H	H	SB	1221	H	SB	1331	J	SB
1042I thru 1043B	G	SB	1222 thru 1223	J	SB	1332	J	SB
1043C thru 1043F	H	SB	1224 thru 1225A	H	SB	1333	I	SB
1043G thru 1044E	G	SB	1226 thru 1227B	K	SB	1334	I	SB
1045 thru 1045F	B	SB	1228 thru 1229A	J	SB	1343 thru 1344A	H	SB
1046 thru 1051B	H	SB	1230 thru 1231	H	SB	1345	J	SB
1052	G	SB	1232	G	SB	1348 and 1348A	G	SB
1053 thru 1054C	H	SB	1233 thru 1237B	K	SB	1350 thru 1350C	I	SB
						1550A	P	UB

Type number-to-letter cross reference chart for use with the following specifications charts

Type No.	Column Letter		Type No.	Column Letter		Type No.	Column Letter	
1351 thru 1351B	K	SB	1427 and 1427A	H	SB	1493 and 1493A	G	UB
1352 thru 1352B	I	SB	1428 thru 1429A	P	SB	1494 and 1495A	B	UB
1353	K	SB	1430A	G	UB	1496	G	UB
1354 thru 1355A	I	SB	1431	P	SB	1497	A	UB
1356 and 1356B	K	SB	1432 and 1432A	G	UB	1498	E	UB
1357 thru 1358	I	SB	1433	K	SB	1499	F	UB
1359 thru 1362B	K	SB	1434 thru 1435A	P	SB			
1363 thru 1369A	G	SB	1436 and 1436A	I	SB			
1372 thru 1375A	G	SB	1437	J	SB			
1376 and 1376A	J	SB	1439	I	SB	1500	E	UB
1377 thru 1378	K	SB	1440 thru 1440D	A	UB	1501A thru 1501E	A	UB
1379 thru 1379A	H	SB	1441	P	SB	1503 thru 1503D	L	UB
1380 thru 1380B	K	SB	1442 thru 1442B	G	UB	1506 thru 1507	F	UB
1381	G	SB	1443	I	SB	1508	G	UB
1382	H	SB	1444 and 1444A	G	UB	1509	C	UB
1383 thru 1383B	K	SB	1445	I	SB	1510	L	UB
1384 thru 1385	G	SB	1446 and 1446A	A	SB	1511	C	UB
1386	I	SB	1447	G	SB	1512 and 1512A	B	UB
1387 thru 1388	K	SB	1448 thru 1450	F	UB	1513	L	UB
1389	G	SB	1450A thru 1450B	F	UB	1515 thru 1516C	C	UB
1390 thru 1390B	H	SB	1450C thru 1450E	F	UB	1517	E	UB
1391	K	SB	1453	K	SB	1518	D	UB
1392	J	SB	1454 and 1454A	A	UB	1519 thru 1521	A	UB
1393	G	SB	1455 thru 1456	K	SB	1522	L	UB
1394 thru 1395A	P	SB	1459	G	UB	1523	A	UB
1396	K	SB	1460 thru 1460F	A	UB	1524	B	UB
1397	H	SB	1461	K	SB	1527	C	UB
1398 thru 1399	K	UB	1462	A	UB	1528	A	UB
			1463	K	SB	1529A and 1529B	C	UB
			1464 thru 1464B	L	UB	1530 thru 1530B	A	UB
			1465	A	UB	1531 thru 1535B	C	UB
1400	K	UB	1466 thru 1466A	F	UB	1536	L	UB
1401 thru 1401F	F	UB	1467 thru 1468	K	SB	1537	A	UB
1402 and 1402B	G	UB	1471 thru 1471B	E	UB	1538 thru 1541A	L	UB
1403A and 1403B	K	SB	1472 thru 1472C	L	UB	1542	E	UB
1404 and 1404A	K	SB	1473 thru 1473B	A	UB	1543 thru 1546	A	UB
1405 thru 1406A	H	SB	1474	L	UB	1547	C	UB
1407 thru 1408	K	SB	1475 thru 1476	A	UB	1549	C	UB
1409A	J	SB	1477	G	SB			
1410 thru 1412A	I	SB	1478	J	SB			
1413 thru 1416	I	SB	1479	G	UB			
1417 and 1417A	K	SB	1482 and 1482A	F	UB	S-1801 thru 1822	F	SB
1418 and 1418A	P	SB	1483	F	UB	1823 and 1824	E	SB
1419 and 1419A	K	SB	1484 thru 1484D	A	UB			
1420 and 1420A	H	SB	1485	G	UB			
1421 and 1421A	K	SB	1486	D	UB			
1422 thru 1423	P	SB	1487	J	SB	2001 thru 2003	B	SB
1424	K	SB	1488 thru 1488D	A	UB	2004 thru 2006	D	SB
1425	G	UB	1489 thru 1490B	C	UB	2007 thru 2007B	B	SB
1426 thru 1426B	P	SB	1491	L	UB	2008 thru 2008B	D	SB

Type number-to-letter cross reference chart for use with the following specifications charts

Type No.	Column Letter		Type No.	Column Letter		Craftsman No.	Column Letter	
2009	B	SB	40053 thru 40054A	N	SB	200.183112	F	UB
2010 thru 2011	D	SB	40056 thru 40060B	O	SB	200.183122	F	UB
2012	F	SB	4006¹ thru 40062B	N	SB	200.193132	F	UB
2013 thru 2014	B	SB	40063 thru 40064	O	SB	200.193142	F	UB
2015 thru 2018	D	SB	40065 thru 40066	N	SB	200.193152	G	UB
2020	F	SB	40067 thru 40068	O	SB	200.193162	G	UB
2021	D	SB	40069 thru 40075	N	SB	200.203172	H	UB
2022 thru 2022B	F	SB				200.203182	H	UB
2023 thru 2026	D	SB				200.203192	H	UB
2027	B	SB				200.213112	H	UB
2028 thru 2029	D	SB	710101 thru 710116	E	SB	200.213122	H	UB
2030 thru 2031	B	SB	710124 thru 710130	E	SB	200.503111	F	UB
2032 thru 2033A	D	SB	710131 thru 710137	E	SB	200.583111	F	UB
2034 thru 2035C	F	SB	710138 thru 710149	E	SB	200.593121	F	UB
2036 thru 2037	B	SB	710154	M	SB	200.613111	F	UB
2038	F	SB	710201 thru 710209	E	SB			
2039 thru 2044	C	SB	710210 thru 710218	E	SB			
2045 thru 2046B	F	SB	710219 thru 710227	E	SB			
2047 thru 2048	D	SB	710228	M	SB			
2049 thru 2049A	F	SB	710150	G	SB			
2050 thru 2050A	D	SB	710151	H	SB			
2051 thru 2052A	D	SB	710155	L	SB			
2053 thru 2055	F	SB	710157	H	SB			
2056	D	SB	710229	H	SB			
2057	F	SB	710230	N	SB			
2058 thru 2058B	D	SB	710234	N	SB			
2059 thru 2063	F	SB						
2064 thru 2064A	D	SB						
2065	F	SB						
2066	D	SB						
2067 thru 2071B	F	SB						
2200 thru 2201A	D	SB						
2202 thru 2204	E	SB						
2205 thru 2205A	D	SB						
2206 thru 2206A	G	SB						
2207	D	SB						
2208	G	SB						
2768	C	SB						
40001 thru 40028	N	SB						
40029 thru 40032C	O	SB						
40033	N	SB						
40034 thru 40045A	O	SB						
40046	N	SB						
40047 thru 40052	O	SB						

Type number-to-letter cross reference chart for use with the following specifications charts

	A	B	C	D	E	F	G	H
Bore	1.500 / 1.5005	1.6253 / 1.6258	1.7503 / 1.7508	1.7503 / 1.7508	2.000	2.000	2.000	2.000
Stroke	1.375	1.50	1.50	1.50	1.50	1.50	1.50	1.50
Displacement Cubic Inches	2.43	3.10	3.60	3.60	4.70	4.70	4.70	4.70
Point Gap	.020	.020	.020	.020	.020	.020	.015	.015
Timing B.T.D.C. Before Top Dead Center	1/16" or .0625	5/32" or .1562	5/32" or .1562	1/4" or .250	5/32" or .1562	5/32" or .1562	5/32"	1/4" or .250 11/64" for "super"
Spark Plug Gap	.030	.030	.030	.030	.030	.030	.030	.030
Piston Ring End Gap	.003 / .008	.005 / .010	.005 / .010	.005 / .010	.006 / .011	.006 / .011	.006 / .011	.006 / .011
Piston Diameter	1.4966 / 1.4969	1.6216 / 1.6219	1.7461 / 1.7464	1.7461 / 1.7464	1.9948 / 1.9951	1.9948 / 1.9951	1.9948 / 1.9951	1.9949 / 1.9955
Piston Ring Groove Width	.095 / .096	.095 / .096	.095 / .096	.095 / .096	.095 / .096	.095 / .096	.095 / .096	.095 / .096
Piston Ring Width	.093 / .0935	.093 / .0935	.093 / .0935	.093 / .0935	.093 / .0935	.093 / .0935	.093 / .0935	.093 / .0935
Piston Pin Diameter	.3750 / .3751	.3750 / .3751	.3750 / .3751	.3750 / .3751	.3750 / .3751	.3750 / .3751	.3750 / .3751	Early .3750 / .3761 Late .4997 / .4999
Connecting Rod Diameter Crank Bearing	.6869 / .6874	.6869 / .6874	.6869 / .6874	.6986 / .6989 w/o needles	.6869 / .6874	.6869 / .6874	.6869 / .6874	.6935 / .6939 w/o needles
Crankshaft Rod Needle Diameter				.0653 / .0655				.0653 / .0655
Crank Pin Journal Diameter	.6860 / .6865	.6860 / .6865	.6860 / .6865	.5615 / .5618	.6860 / .6865	.6860 / .6865	.6860 / .6865	.5615 / .5618
Crankshaft P.T.O. Side Main Brg. Dia.	.6689 / .6693	.6689 / .6693	.6690 / .6694	.6689 / .6693	.9995 / 1.0000	.6689 / .6693	.9995 / 1.0000	.9995 / 1.0000
Crankshaft Magneto Side Main Brg. Dia.	.6689 / .6693	.6689 / .6693	.6689 / .6693	.6690 / .6694	.7495 / .7500	.6689 / .6693	.7495 / .7500	.7495 / .7500
Crankshaft End Play	.003 / .008	.003 / .008	.003 / .008	.003 / .008	.009 / .022	.009 / .022	.009 / .022	.009 / .022

Specifications for 2 cycle engines with split crankcases

I	J	K	L	M	N	O	P		
2.000	2.000	2.093 / 2.094	2.2505 / 2.2510	2.2505 / 2.2510	2.5030 / 2.5035	2.5030 / 2.5035	2.000		Bore
1.50	1.625	1.63	2.00	2.00	1.625	1.680	1.50		Stroke
4.70	5.10	5.80	8.00	8.00	7.98	8.25	4.70		Displacement Cubic Inches
.015	.020	.015	.020	.020	.020	.020	.015		Point Gap
11/64" or .175	11/64" or .175	3/32" or .095	1/8" or .125	3/32" or .095	11/64" or .175	11/64" or .175	.90		Timing B.T.D.C. Before Top Dead Center
.035	.030	.030	.030	.030	.030	.030	.035		Spark Plug Gap
.006 / .011	.006 / .011	.006 / .011	.007 / .015	.007 / .015	.005 / .013	.005 / .013	.006 / .011		Piston Ring End Gap
1.9948 / 1.9951	1.9951 / 1.9948	2.0880 / 2.0883	2.2460 / 2.2463	2.2460 / 2.2463	2.4960 / 2.4963	2.4960 / 2.4963	1.9948 / 1.9951		Piston Diameter
.095 / .096	.095 / .096	T.0655 / .0665 L.0645 / .0655	.095 / .096	.095 / .096	T.0655 / .0665 L.0645 / .0655	T.0655 / .0665 L.0645 / .0655	.095 / .096		Piston Ring Groove Width
.093 / .0935	.093 / .0935	.0615 / .0625	.093 / .0935	.093 / .0935	.0615 / .0625	.0615 / .0625	.093 / .0935		Piston Ring Width
.4997 / .4999	.3750 / .3751	.4997 / .4999	.5000 / .5001	.3000 / .5001	.4997 / .4999	.4997 / .4999	.4997 / .4999		Piston Pin Diameter
.6941 / .6944	.6869 / .6874	.9407 / .9412	1.000 / 1.0004	1.000 / 1.0004	.9407 / .9412	.9407 / .9412	.6941 / .6944		Connecting Rod Diameter Crank Bearing
.0653 / .0655		.0943 / .0945	.0943 / .0945	.0943 / .0945	.0943 / .0945	.0943 / .0945	.0653 / .0655		Crankshaft Rod Needle Diameter
.6860 / .6865	.6860	.7499 / .7502	.8096 / .8099	.8096 / .8099	.7499 / .7502	.7499 / .7502	.6860 / .6865		Crank Pin Journal Diameter
.9995 / 1.0000	.9995 / 1.0000	.6990 / .6994	.9839 / .9842	.9839 / .9842	.7871 / .7875	.7871 / .7875	.9995 / 1.0000		Crankshaft P. T. O. Side Main Brg. Dia.
.7495 / .7500	.7495 / .7500	.6990 / .6994	.9839 / .9842	.9839 / .9842	.7498 / .7501	.7498 / .7501	.7495 / .7500		Crankshaft Magneto Side Main Brg. Dia.
.009 / .022	.009 / .022	.003 / .008	.003 / .008	.003 / .008	.008 / .013	.008 / .013	.009 / .022		Crankshaft End Play

Specifications for 2 cycle engines with split crankcases

	A	B	C	D	E	F**	G	*H
Bore	2.093/2.094	2.093/2.094	2.093/2.094	2.093/2.094	2.093/2.094	2.093/2.094	2.093/2.094	2.093/2.094
Stroke	1.250	1.41	1.41	1.41	1.41	1.50	1.50	1.50
Displacement Cubic Inches	4.3028	4.8536	4.8536	4.8536	4.8536	5.1632	5.1632	5.1558
Point Gap	.015	.015	.015	.015	.015	.020	.015	.020
Timing B.T.D.C. Before Top Dead Center	.150	.100	.140	.100	.140	.100	.185	.085
Spark Plug Gap	.035	.030	.035	.035	.035	.030	.035	.035
Piston Ring End Gap	.006/.011	.006/.011	.006/.011	.006/.011	.006/.011	.006/.014	.006/.011	.006/.011
Piston Diameter	2.0880/2.0883	2.0880/2.0883	2.0880/2.0883	2.0880/2.0883	2.0880/2.0883	2.0880/2.0883	2.0880/2.0883	2.078/2.081
Piston Ring Groove Width	.0645/.0665	.0645/.0665	.0645/.0665	.0645/.0665	.0645/.0665	top .0980/bot .0960	.0645/.0665	.0645/.0655
Piston Ring Width	.0615/.0625	.0615/.0625	.0615/.0625	.0615/.0625	.0615/.0625	.0925/.0935	.0615/.0625	.0615/.0625
Piston Pin Diameter	.4997/.4999	.3750/.3751	.4997/.4999	.37500/.37515	.4997/.4999	.3750/.37515	.4997/.4999	.4997/.4999
Connecting Rod Diameter Crank Bearing	.6941/.6944	.6876/.6880	.6941/.6944	.6876/.6880	.6941/.6944	.6876/.6880	.6941/.6944	1.0443/1.0438 without liner & bearings
Crankshaft Rod Needle Diameter	.0653/.0655		.0652/.0655		.0653/.0655		.0653/.0655	.0780/.0781
Crank Pin Journal Diameter	.5614/.5618	.6860/.6865	.5615/.5618	.6860/.6865	.5615/.5618	.6860/.6865	.5614/.5618	.8425/.8430
Crankshaft P.T.O. Side Main Brg. Dia.	.6991/.6695	.6695/.6691	.6695/.6691	.6695/.6691	.6695/.6691	.8750/.8745	.6691/.6695	.9995/1.0000
Crankshaft Magneto Side Main Brg. Dia.	.6691/.6695	.6691/.6695	.6691/.6695	.6691/.6695	.6691/.6695	.7495/.7500	.6691/.6695	.7495/.7500
Crankshaft End Play	.0152/.0172	NONE	NONE	NONE	NONE	.004/.018	.0152/.0172	.0025/.0157
Compression Pressure At Cranking Speeds	90 lbs.	90 lbs.	90 lbs.	90 lbs.	90 lbs.	90 lbs.	90 lbs.	90 lbs.

Specifications for 2 cycle engines without split crankcases

*I	**J	*K	L	**M	***N	***O		
2.093/2.094	2.093/2.094	2.093/2.094	2.375/2.376	2.375/2.376	2.093/2.094	2.437/2.438		Bore
1.50	1.76	1.75	1.0875	1.71	1.50	1.75		Stroke
5.1558	6.0494	6.00	7.4453	7.503	5.13	8.17		Displacement Cubic Inches
.020	.020	.020	.015	.020	.018	.020		Point Gap
.085	.085	.100	.115	.095	.100	.100		Timing B.T.D.C. Before Top Dead Center
.035	.035	.035	.035	.035	.035	.035		Spark Plug Gap
.006/.011	.006/.011	.006/.014	.005/.013	Both .005/.013	.006/.014	.007/.017		Piston Ring End Gap
2.078/2.081	2.078/2.081	2.0880/2.0883	2.3700/2.3703	2.360/2.363	2.0880/2.0883	2.4320/2.4310		Piston Diameter
.0645/.0655	.0645/.0655	.0955/.0985	.0655/.0665	top .0655/bot. .0665	top .0980/bot. .0960	top .0660/bot. .0650		Piston Ring Groove Width
.0615/.0625	.0615/.0625	.0925/.0935	.0110/.0700	Both .0615/.0625	Both .0925/.0935	Both .0615/.0625		Piston Ring Width
.4997/.4999	.4997/.4999	.4997/.4999	.4997/.4999	.4997/.4999	.3750/.3751	.4997/.4999		Piston Pin Diameter
.525/.540	.525/.540	.6876/.6880	.7566/.7569	.7566/.7569	.6876/.6880	.7588/.7592		Connecting Rod Diameter Crank Bearing
			.0653/.0655	.0653/.0655		.0653/.0655		Crankshaft Rod Needle Diameter
.6860/.6865	.6860/.6865	.6860/.6865	.6240/.6243	.6239/.6243	.6860/.6865	.6262/.6266		Crank Pin Journal Diameter
.8745/.8750	.8745/.8750	.8745/.8750	.6691/.6695	.6691/.6695	.8745/.8750	.6695/.6691		Crankshaft P.T.O. Side Main Brg. Dia.
.7495/.7500	.7495/.7500	.7495/.7500	.6691/.6695	.7495/.7500	.7495/.7500	.8745/.8750		Crankshaft Magneto Side Main Brg. Dia.
.0025/.0157	.0025/.0157	.004/.018	.0152/.0172	.0081/.0263	.0025/.0157	NONE		Crankshaft End Play

Specifications for 2 cycle engines without split crankcases

Application	Size	Torque
CARBURETOR AND REED PLATES		
Carburetor to crankcase, Carburetor to adapter, or Carburetor adapter to crankcase	1/4 in.-20 hex nuts 1/4 in.-20 x 5/16 hex hd. cap screws	70-75 in. pounds
Carburetor to snow blowers cover	1/4-28 center lock hex nut	30-35 in. pounds
Carburetor outlet fitting	1/8 pipe plug	40 in. pounds
Reed and cover plates 639 type engines	1/4-28 x 1-1/4	50-60 in. pounds with Loctite, type A
Reed to plate 635 type engines	6-32 x 3/8 fil. flex.	12-18 in. pounds
CRANKCASE AND CYLINDER		
Crankcase to crankcase cover	10-24 x 5/8 or 1/2 fil hd Sems	23-30 in. pounds
Crankcase to crankcase cover screws	10-24 x 5/8 or 3/4 fil flex hd	35-40 in. pounds
Mounting cylinder or carburetor to crankcase studs	1/4-20 studs	50 in. pounds
Base to crankcase	5/16-24 nut (650482) to stud (650471)	240-250 in. pounds
Cylinder to crankcase nuts	1/4-20 hex nuts	70-75 in. pounds
In cylinder	Pipe plug (650414)	100-110 in. pounds
Spark plug stop lever and head shroud to head	1/4-20 x 5/8 pan flex hd	50-60 in. pounds
Cylinder head to cylinder	10-24 x 3/4 pan hd mach screw	30-40 in. pounds
Cylinder head to cylinder	10-24 x 3/4 socket hd cap 1/4-20 pan flex head	50-60 in. pounds 80-90 in. pounds
Cylinder head to cylinder (635 type engines)	10-24 x 7/8 pan flex hd	45-50 in. pounds
Cable clip and transfer port cover to cylinder	10-24 x 5/8 in. fil hd Sems	25-30 in. pounds
Cable clip to cylinder Stop lever to cylinder	10-24 x 1/2 fil hd Sems	25-30 in. pounds
Spark plug		18-22 ft pounds

Torque specifications for 2 cycle engines

Application	Size	Torque
CRANKCASE AND CYLINDER (Cont.)		
Transfer cover cross port engines	10-24 x 1/2 pan flex hd	25-30 in. pounds
CRANKSHAFT AND CONNECTING RODS		
Aluminum and bronze rods to rod cap	10-24 x 17/32 fil hd screws	40-50 in. pounds
Steel rod to rod cap	10-32 x 9/16 and 5/8 soc hd screws	70-80 in. pounds
Flywheel nut on tapered end of crankshaft: Aluminum hub on iron or steel shaft.	7/16-20 RH or LH hex nut	18-25 ft pounds
Steel hub flywheel on iron shaft	7/16-20 hex nut	18-25 ft pounds
Steel hub flywheel on steel shaft	7/16-20 hex nut	30 ft pounds

Torque specifications for 2 cycle engines

4 · Lawn-Boy

Engine Identification

All Lawn-Boy engines are of 2 cycle design. The two series, the D-400 and the D-600, can be recognized by their different cylinder head design. The D-600 engine's cylinder head has an "H" shaped casting which looks like it might be a brace for the outside cooling fins. The D-400 engine's cylinder head has just one brace-like casting.

Carburetion

The carburetors used on Lawn-Boy engines are float feed type. The same servicing and rebuilding procedures can be used for both D-400 and D-600 carburetors, however there are some differences which will be covered in the following text.

Engine identification

D-400 SERIES CARBURETORS

Adjustment

1. To adjust the needle of the main nozzle, turn the adjusting knob clockwise until the needle seats lightly. Do not force the needle down too tight, since you may damage the needle or the seat in the nozzle and make further adjustment difficult.

2. Open the needle two turns and start the engine. If the engine begins to die, open the carburetor adjusting knob further by turning it counterclockwise. If the engine runs roughly, close the knob by turning it clockwise until the engine runs smoothly.

3. Let the engine run for about five minutes to warm up. Close the adjusting knob slowly until the engine begins to die, then open the adjusting knob ¼ to ½ of a turn. The carburetor is now properly adjusted.

4. Early D-400 carburetors also have a slow speed adjustment. The correct setting for this adjustment is one turn open from completely closed.

5. If the engine runs roughly or hunts at low speeds, the condition can be corrected by either opening or closing the slow speed adjusment from the original setting of one turn open.

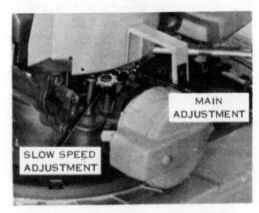

Early D-400 series carburetor adjustments

6. On newer D-400 engines, the low speed adjusting valve is replaced by a fixed jet and a lead filled hole.

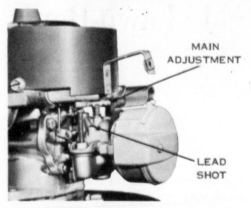

MAIN
ADJUSTMENT

LEAD
SHOT

Late D-400 series carburetor adjustments

7. The float is set by removing the float bowl and gasket and inverting the carburetor. With the float arm resting on the float valve needle, the top of the float should be $^{15}/_{32}$ in. above the edge of the carburetor body.

8. To adjust the float, bend the metal arm with a pair of needle nose pliers. Do not attempt to bend the float arm by pressing on the cork float.

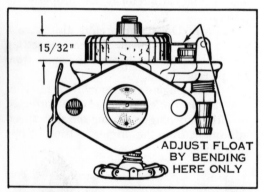

15/32"

ADJUST FLOAT
BY BENDING
HERE ONLY

Float adjustment for all D-400 and D-600 carburetors

Servicing

Servicing the carburetor consists of removing it from the engine, cleaning it, inspecting and replacing worn or damaged parts, and making adjustments.

1. To remove the carburetor from the engine, the carburetor and reed plate must be removed as an assembly. The reed plate is then removed from the carburetor.

NOTE: *Carburetor servicing kits include two self-threading screws. The reed plate mounting flange on the carburetor is drilled but not tapped. These screws form a very tight fit, eliminating the possibility of the screw vibrating loose during engine operation.*

2. All parts, except the cork float, can be cleaned in carburetor cleaner and blown dry with compressed air. A strong solvent will remove the sealer from the float. Carburetor parts should not be dried with a cloth because of the possibility of lint sticking to the parts.

3. Inspect the valve needles for grooves or other evidence of wear or damage. Replace the needles if necessary. The manufacturer recommends that a magnifying glass be used in the inspection.

4. The reed plate can be cleaned with carburetor cleaner. Be careful when handling the reeds so as not to distort them. The reeds must lie flat against the reed plate to form a perfect seal. Bent or damaged reeds cannot be repaired.

5. On A Series engines, the complete reed plate assembly must be replaced.

NOTE: *Do not use compressed air on the reeds.*

6. The rough edge of the reed should be away from the plate. Maximum clearance allowed between the reed and the reed plate is 0.015 in. If the clearance exceeds this, the reeds must be replaced.

7. When reasembling the carburetor, use new gaskets. When installing the nozzle in the carburetor, unscrew the control knob a few turns to avoid accidentally tightening the needle on the seat and damaging it. Use new lock nuts, if the nylon material on the old carburetor is excessively worn, to ensure a tight seal between the reed plate and the crankcase.

8. Check the rubber washer of the primer and replace it if cracked. Use the new primer spring that comes with the repair kit.

9. There is a fuel nozzle filter installed on the main fuel nozzle to minimize the possibility of dirt getting into the jet area and causing damage and stalling. Firmly install the filter on the nozzle.

10. Hold the carburetor upside down so that the float inlet needle is in the closed position; carefully place the fuel bowl in position.

11. The throttle plate is installed in the carburetor with the two half moons embossed in the surface of the plate facing toward the reed valve assembly.

12. The throttle shaft is split for mount-

ing the throttle plate. Check the $\frac{7}{32}$ in. diameter shaft for wear. When installing the throttle shaft spring, hook the end in the carburetor body and add tension by turning the spring $\frac{1}{2}$ to $\frac{3}{4}$ of a turn and hooking the other end on the throttle shaft arm.

D-600 SERIES CARBURETORS

The carburetors used with the solid state, capacitive discharge ignition engines are considered to be automatic. There is no adjustment to regulate fuel intake. However, there is an air adjustment which only has to be adjusted when there is a significant altitude change. This is, in effect, an atmospheric pressure adjustment that would have to be made if the engine is operating in very high or very low altitudes.

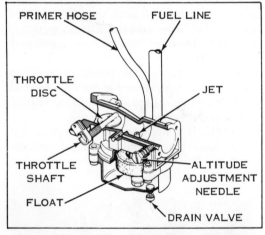

Cutaway diagram of a D-600 series carburetor

PRIMER HOSE FUEL LINE

THROTTLE DISC JET

THROTTLE SHAFT ALTITUDE ADJUSTMENT NEEDLE

FLOAT

DRAIN VALVE

Air Adjustment

1. Normal setting for most geographical areas is usually $\frac{3}{4}$ of a turn open from a closed position. With this setting, the correct amount of air enters the carburetor bowl vent allowing the right amount of fuel to mix with the air entering the carburetor venturi.

2. At low altitudes, the air adjusting needle may have to be opened slightly to allow more fuel to mix with the heavier incoming air.

3. At high altitudes, the air adjusting needle may have to be closed slightly to reduce the amount of fuel that is to be mixed with the lighter incoming air.

4. If the air adjustment is in the completely closed position, no air is entering the carburetor vent passage. Therefore, the float bowl pressure has been eliminated and the fuel supply to the carburetor venturi is cut off.

Service

D-600 carburetors are serviced in the same manner as D-400 carburetors.

Governor

Adjustment

1. In order to adjust the governor, the flywheel has to be removed.

2. Before making the main adjustment, turn the variable speed knob to the "light" position and rotate the governor thrust collar clockwise, as far as it will go.

SPRING GOES INSIDE GOVERNOR COLLAR

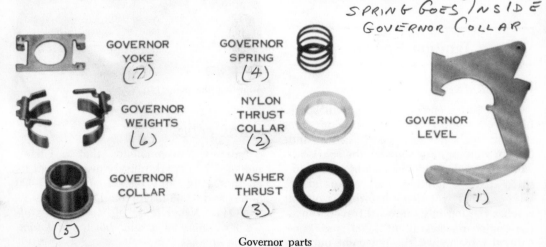

GOVERNOR YOKE (7)

GOVERNOR SPRING (4)

GOVERNOR WEIGHTS (6)

NYLON THRUST COLLAR (2)

GOVERNOR LEVEL

GOVERNOR COLLAR (5)

WASHER THRUST (3)

(1)

Governor parts

NUMBERED IN ORDER OF ASSEBLY

3. Press down firmly on the governor assembly and depress the throttle shaft to the closed position. If the governor is properly adjusted, there should be about $\frac{1}{16}$ in. between the top of the governor rod and the bottom of the governor lever.

4. To make any necessary adjustments, simply bend the governor lever up or down with a pair of pliers to obtain proper clearance.

NOTE: *The governor lever is diagonally creased to simplify adjustment. Bend the lever along the crease.*

Governor action is continuous and there should be no noticeable variation in engine speed.

SPECIAL TOOL

GOVERNOR THRUST WASHER

1/16"

GOVERNOR ROD

THROTTLE SHAFT

Governor adjustment

Ignition System

D-400 SERIES

D-400 series engines have a twin spark ignition system. This system provides two different spark timings, one for starting and one for running. For starting, the spark-advance flyweight holds the cam in such a position that the igniting spark occurs at 6° (of crankshaft rotation) before the piston reaches the top of its upward travel. When the engine reaches about 1,000 rpm, centrifugal force moves the flyweight out, ro-

tating the cam to such a position that the ignition spark now occurs at 26° (of crankshaft rotation) before the piston reaches the top of its upward travel.

Flywheel-to-Coil Adjustment

The air gap between the coil heels and the flywheel magnets should be 0.010 in. To check the gap, it is best to insert a strip of 0.010 in. non-metallic shim stock between the coil heels and the flywheel magnets. The gap is adjusted by loosening the coil mounting screws.

Breaker Point Adjustment

1. The point gap should be 0.020 in. To check the gap, rotate the crankshaft until the wear block is centered on the lobe of the cam. Move the crankshaft toward the carburetor and hold it in that position.

2. Loosen the breaker base screw and place the gauge between the points. Pivot the breaker base until the gap is correct.

3. Retighten the breaker base screw and recheck the gap.

4. Check the breaker points every 40 to 50 hours of operation for wear or pitted contact surfaces. Replace the points if they are pitted or worn.

NOTE: *It is not necessary to replace the condenser every time the points are replaced. The condenser will usually last the life of the engine.*

Adjusting the point gap

SPARK PLUG

Lawn-Boy recommends that a Champion® CJ-14 spark plug be used for all its engines. Plug gap is 0.025 in. for the D-400 Series and 0.035 in. for the D-600 Series.

NOTE: *Never reuse old spark plugs after cleaning in a sand blaster in Lawn-Boy engines.*

D-600 SERIES

D-600 engines have a solid state ignition system which uses electronic devices such as diodes, transistors, silicon controlled rectifiers, and other semiconductors to take the place of one or more standard ignition components.

This system is called a C-D (capacitive discharge) system and has no moving parts other than the permanent magnets located in the flywheel.

There is very little that can be done with this type of system as far as troubleshooting is concerned. If there is a problem with this system, there are only five checks you can make: ignition switch, switch lead, flywheel, air gap between the C-D pack and the flywheel, and the spark plug. The only troubleshooting procedure is to see if the system is producing a spark.

NOTE: *The most vulnerable part of the ignition system is the ignition switch.*

CAUTION: *If the engine has been running, allow 10 seconds before removing the spark plug lead wire. This will allow the charge in the C-D pack to leak off.*

Engine Mechanical

Because of the inherent simplicity of these engines, only disassembly and reasasembly procedures will be given. Refer to the specifications chart at the end of this section for torque specifications, measurements, and clearances.

Disassembly and Reassembly

D-400

1. Disconnect the gas line from the carburetor and remove the complete engine shroud, gas tank, gas tank bracket, primer bar, and air baffle.

NOTE: *Replace each shroud screw in its original hole.*

2. Remove the spark plug and install a piston stop tool. This will allow easy removal of the flywheel nut by not allowing the crankshaft to turn. Remove the flywheel nut and trim plate. On some models, the trim plate must be removed to expose the flywheel nut.

3. Remove the flywheel by hitting the

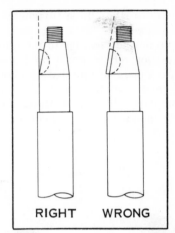

RIGHT   WRONG

Be sure that the key is installed correctly

top of the crankshaft with a soft mallet while simultaneously placing pressure on the underside of the flywheel. Check the flywheel carefully for cracks and enlargement of the keyway.

4. Remove the flywheel key. Use a pair of pliers to roll the key out of the crankshaft keyway.

5. Lift the governor yoke, weights, and collar off as an assembly. Set it aside carefully.

NOTE: *Reinstall the governor assembly with the word "KEY" on the yoke facing the crankshaft keyway.*

6. Remove the governor spring and steel thrust washer.

7. Unhook the variable speed spring from the governor lever. Lift up on the governor lever to disengage the prongs from the slots in the dust cover. Remove the governor lever and the nylon thrust collar.

8. Remove the dust cover by taking out the three attaching screws.

9. Push the small end of the spark advance flyweight toward the crankshaft. Hold the tension of the spring against the crankshaft and return the flyweight to its original position. Allow the smaller end of the flyweight to drop down, then remove the pin and spring.

NOTE: *Install the flyweight with the words "SHAFT KEYWAY THIS END" (written on the small end) toward the crankshaft keyway. Reinstall the pin and spring properly with the clip on the end installed horizontally in the recess in the narrow end of the flyweight.*

10. Slide the flyweight and cam off of the crankshaft.

11. Remove the three magneto plate mounting screws.

12. The needle bearings in the armature plate are not retained. To hold them in place, use a piece of round stock about the same diameter as the crankshaft. As the armature plate is lifted from the crankshaft, insert the round stock in place of the crankshaft.

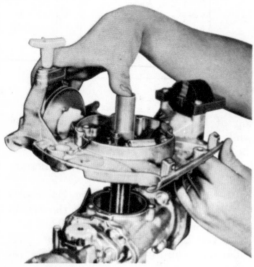

Removing the magneto plate

13. To replace the bearing, remove the seal by inserting a screwdriver under the seal and prying up. Holding the armature plate in your hand, drive out the old bearing using a suitable driver and a soft mallet. Place the new bearing in place on the armature plate with the lettering facing up, then drive the bearing into its seat using a suitable installer tool and a soft mallet. Do this while still holding the plate in your hand; otherwise, the armature plate might be broken. Install a new seal in the same manner.

14. Remove the carburetor and reed plate assembly together by removing the four screws which hold the reed plate to the crankcase.

NOTE: *The upper left-hand screw has no lock washer. This is to provide clearance between the throttle arm and the head of the screw.*

15. Use a large screwdriver to loosen, but not remove, the connecting rod screws. The lock tabs do not have to be bent away from the screws first.

16. Remove the four ⅜ in. hex head screws which hold the cylinder to the crank-

case. Hit the cylinder head with a soft-headed hammer to break it loose. Remove the cylinder by pulling it quickly away from the piston. This prevents the rings from binding and/or breaking.

17. Make sure that you install the cylinder gasket correctly and not partially blocking the intake ports.

INCORRECT   CORRECT

Installation of the cylinder gasket

18. Position the crankcase so that the piston dome is facing up. The crankshaft should be rotated so that the connecting rod journal is at its lowest position. Remove the rod cap, allowing the needle bearings to fall out. Count the needles. There should be 33.

NOTE: *Install new lock tabs on the rod cap screw for reassembly.*

19. Remove the piston and rod assembly. Note the alignment marks on the rod and cap and the dovetail ends of the bearing inserts. These parts must be mated for reassembly.

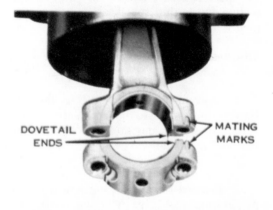

DOVETAIL ENDS   MATING MARKS

Alignment marks on the connecting rod and cap

20. Remove the wrist pin retainer and rings and drive the wrist pin out.

NOTE: *Engines produced prior to 1970 contained three-ring pistons and those engines built since then have two-ringed pistons.*

The two-ring pistons and rod assemblies have 27 loose needles in the rod. Be careful when removing the wrist pin so that loose needles are not misplaced or lost. A $7/16$ in. diameter x $3/4$ in. dowel rod may be inserted to prevent this. This is most important because replacement bearings are not available.

When replacing the retainer rings, the opening must face the piston dome or the piston skirt. This will prevent the retainer rings from popping out during installation. The retainer has a bevelled side and a flat side and, when installed, the flat side must face to the outside of the piston. The word "TOP" is cast into the skirt of the piston and must face up when installed.

21. Pull the crankshaft from the crankcase and wipe dry the connecting rod throw. Apply a coating of heavy grease to the cap and place 16 of the needles on the rod and 17 of the needles on the rod cap.

22. Install the crankshaft in the crankcase. Apply oil to the piston, rings, wrist pin, and cylinder sleeve. Place the rod on the crankshaft journal and install the rod cap. Tighten the rod cap screw finger tight, for the time being.

23. Install a ring compressor over the head of the piston and compress the rings. Place the cylinder over the piston head. Maintain pressure on the ring compressor until the rings enter the cylinder.

NOTE: *Check the cylinder gasket for proper installation.*

24. Center the rod on the wrist pin and tighten the cap screws to the specified torque.

25. Bend the lock tabs over the screws. If the tabs do not fit flat against the screws, do not turn the screws, just bend the tab around the corner of the screw head.

26. Assemble the rest of the engine in the reverse order of disassembly.

D-600

1. Remove the high tension lead from the spark plug.

2. Place the fuel shut-off valve in the "OFF" position. Disconnect the fuel and primer hoses from the carburetor. Remove the seven screws which secure the shroud and gas tank assembly and remove the shroud.

3. Remove the spark plug and install a piston stop tool. This stop will prevent the piston from moving, thus holding the crankshaft for easier removal of the flywheel nut. Remove the flywheel nut.

4. Remove the flywheel by striking the top of the crankshaft with a soft mallet and placing pressure on the underside of the flywheel screen. Examine the flywheel keyway for cracks and enlargement.

5. Remove the flywheel key. Use a pair of pliers to roll the key out of the crankshaft keyway.

6. Lift the governor yoke, weights, and collar off as an assembly. Set it aside carefully.

NOTE: *Reinstall the governor assembly with the word "KEY" on the yoke facing the crankshaft keyway.*

7. Remove the governor spring and steel thrust washer.

8. Remove the three magneto plate retaining screws.

9. Unhook the variable speed spring from the governor lever. Lift up on the governor lever to disengage the prongs from the slots in the bracket. Remove the governor lever and the nylon thrust collar.

10. To prevent the loose needle bearings from falling out when the magneto plate is removed, place a length of round stock (with the same outside diameter as the crankshaft) in the opening as the crankshaft is removed and as the magneto plate is lifted off of the engine. The oil seal can be removed by prying it out with the tip of a screwdriver. Once the seal is removed, it must be replaced with a new one.

11. To replace the bearing, drive the old bearing out with a suitable driver while holding the magneto plate in your hand. Install the new bearing with the lettering facing up. Drive the bearing in until the tool bottoms on the plate. This will recess the bearing slightly. Install the new seal. This operation is also performed while holding the magneto plate in your hand because the plate might crack if it is hammered while resting on a hard surface.

12. Remove the carburetor and reed plate assembly together by removing the four screws which hold the reed plate to the crankcase.

13. Use a large screwdriver to loosen the connecting rod screws, but do not remove them.

14. Remove the four ⅜ in. hex head screws which hold the cylinder to the crankcase. Hit the cylinder head sharply with a soft-headed mallet.

15. Remove the cylinder by quickly pulling it away from the piston. This prevents the rings from binding and/or breaking.

16. Examine the cylinder gasket. Make sure that it is installed correctly for reassembly and that it is not blocking the intake ports.

17. Rotate the crankshaft until the rod cap is at its lowest position. The rod cap may now be removed, allowing the needle bearings to fall out. There should be 33 needles. Examine them carefully to make sure that they are not worn and do not have flat spots.

18. Remove the piston and rod assembly. Note the alignment marks on the rod and cap and the dovetail ends of the bearing inserts. These parts must be mated accordingly when reassembled.

19. Remove the wrist pin retainer rings and drive out the wrist pin. The connecting rod has 27 loose needles and care must be taken not to lose any of the loose bearings. A ⁷⁄₁₆ in. diameter by ¾ in. dowel rod may be inserted to prevent this. Bearing replacements are not available. They must be purchased as part of the rod and bearing assembly.

NOTE: *The retainer ring opening must face the piston dome or skirt. This will prevent the rings from popping out during reassembly. The retainer has a bevelled side and a flat side. The flat side should face outward and away from the piston.*

20. The word "TOP" is cast into the skirt of the piston. When installed in the cylinder, "TOP" must face up.

21. Pull the crankshaft from the crankcase and wipe the connecting rod throw dry. Apply a coating of heavy lubricant on the cap and rod journals and place 17 needles on the rod cap and 16 on the rod.

22. Install the crankshaft in the crankcase. Reinstall the thrust washer in the crankcase of a self propelled engine. Apply oil to the piston rings, wrist pin, and the cylinder sleeve. Now gently place the rod on the crankshaft journal and install the rod cap. Install the new lock tabs on the rod cap screws and tighten the screws just

enough to retain the needle bearings. Do not bend the lock tabs at this time.

23. Stagger the ring gaps over the top of the piston, install a ring compressor over the head of the piston, and compress the rings. Place the cylinder over the piston head. Maintain pressure on the ring compressor until the rings enter the cylinder.

NOTE: *Make sure that the cylinder gasket is installed correctly.*

24. Center the rod on the wrist pin. Tighten the connecting rod cap screws to the correct torque.

25. Bend the lock tabs over the screws. If the tabs do not fit flush against the heads of the screws, do not tighten the screws; instead, bend the tab around the head of the screw.

Starters

RECOIL TYPE

Removal

D-400

1. Remove the engine shroud.

2. Remove the starter rope handle and let the rope gradually recoil through the hole in the top of the armature plate.

3. Remove the starter mounting bolt on the underside of the armature plate and take out the complete starter assembly. Hold the assembly together, so that the recoil spring is not accidentally released.

4. To replace the rope, remove the assembly from the engine and remove the push-on retainer on the end of the worm gear.

  a. Slide off the nylon starter pinion. Note that the pinion gear mounts on the worm with the grooved side toward the pulley.

  b. Remove the three screws which hold the pulley plate to the pulley.

  c. Fit the new starter rope into the grooves on the pulley. Replace the pulley plate and tighten the three screws securely.

  d. Hold the starter assembly so that the worm gear points toward you and wind the rope clockwise on the pulley.

NOTE: *Never lubricate the worm gear or the pinion.*

5. To replace the starter spring, position the curled end of the spring on the starter pulley with the end behind the tab.

　　a. Place the cover over the pulley so that the spring is guided through the slot in the cover.

　　b. Clamp the starter in a vise with wooden jaws.

　　c. Wind the starter rope on the pulley and pull the rope to turn the pulley. Turning the pulley will draw the spring into the cover.

　　d. Rewind the pulley and repeat the operation until the end of the spring is hooked into the slot.

D-600

1. Pull the starter rope out about two feet and tie a slip knot next to the rope guide hole. Untie the knot which secures the starter handle. This may require the use of two pairs of pliers. If you have to cut the rope, melt the end with a match to stop fraying. Remove the starter handle. Pull the slip knot out of the rope and allow the rope to rewind on the pulley.

2. To loosen the $\frac{7}{16}$ in. screw which holds the starter assembly to the armature plate, the carburetor and reed plate assembly must be removed from the crankcase. This is necessary because of the closeness of the screw to the reed plate.

3. Remove the starter assembly from the armature plate by grasping it between the thumb and fingers of both hands. Hold the starter assembly firmly together with both hands to keep the spring loaded cup and pin assembly from separating from the starter pulley. The starter spring is loaded and is potentially harmful if released. Place the starter assembly on a bench with the flat side down.

To disassemble the starter:

4. With the starter on the bench, flat side down, remove the retainer ring with a screwdriver. Remove the pinion gear from the worm gear.

NOTE: *Never lubricate the worm gear.*

5. Remove the three self tapping screws from the starter pulley and remove the retainer plate.

6. There is still tension on the spring. To release the tension, hold the two halves to-

Exploded view of a D-600 recoil starter

gether, pointing the shaft downward. Now drop the assembly on the floor. The halves will separate and release the spring.

NOTE: *Apply a light coating of grease before reinstalling the spring.*

Assembly

D-400

1. Place the starter rope retainer on the outside of the starter cup and insert the starter into the armature plate with the protruding end of the starter spring pointed toward the top.

2. Make sure that the end of the starter spring is positioned in the cut-out at the top of the armature plate. One of the prongs on the starter pinion spring should be above the armature plate ledge and the other below it.

3. The starter rope retainer should be locked in place when the starter is positioned correctly in the armature plate.

4. With the starter assembly in position, install the mounting bolt and tighten it securely. Completely wind the rope around the starter pulley and place 1 to $1\frac{1}{2}$ extra turns on the rope for proper tension. Thread the rope through the hole in the armature plate and reinstall the starter handle.

NOTE: *When installing a new starter pinion spring, it should be stretched only far enough to allow it to snap into the groove of the nylon starter pinion. The prongs of the starter pinion spring should*

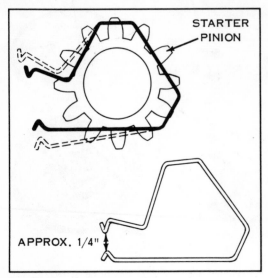

STARTER PINION

APPROX. 1/4"

Installation of the starter pinion spring

be about ¼ in. or less apart when the spring is off the starter pinion.

D-600

1. Anchor the spring in the anchoring channel of the worm gear plate and push the pin of the pin and cup assembly through the worm gear plate opening. Make sure that the spring is guided through the slot opening in the cover.

2. Grasp the pin and cup assembly and wrap your fingers over the worm gear plate to hold the two pieces together. This will keep the spring contained so that you can rotate the worm gear in a clockwise manner to wind the spring. Continue winding until the spring is drawn into the cover and the end hook rests in the recess. Carefully let the spring unwind slowly until the tension is released.

3. Be sure that the sharp bevelled edge of the retainer plate is away from the rope. Reanchor the rope, put the rope retainer plate on, and rewind the rope clockwise. Replace the pinion gear and retainer. The starter assembly is now ready to install on the engine.

4. Position the starter spring so that one prong goes above the armature plate ledge and the other below.

5. Before tightening the screw which secures the starter assembly, make sure that the spring end hook faces up and locates in the die cast recess in the armature plate.

6. Mark a spot on the starter pulley and pull enough rope out so that the mark rotates approximately two full turns. Grasp the starter pulley and spring housing firmly to keep the pulley from returning to its relaxed position and wind the excess rope around the starter pulley.

7. Thread the starter rope up through the rope guide opening, pull approximately 1½ feet of rope through, and tie a knot in it so that it will not slip back. Now, install the handle, remove the slip knot, and allow the starter to rewind until the handle seats.

Engine Specifications

MODELS	A-10, A-11, A-13	C-10, C-12, C-20, C-21, C-22, C-70, C-71, C-72, C-73, C-75	C-13 thru C-17 C-40 thru C-42 C-50, C-51, C-60, C-61, C-80, C-81, C-74, C-76 thru C-78	C-18, C-19, C-19B, C-43, C-44 D-400 thru D-408, D-405E thru D-408E, D-430, D-431, D-440 thru D-448 D-445E thru D-448E D-450 thru D-452 D-460 thru D-462 D-475 D-600
Horsepower			2.5 (3 H.P.)	3½
Bore	1¾ in.	1¹⁵⁄₁₆ in.	2⅛ in.	2⅜
Stroke	1½ in.	1½ in.	1½ in.	1½ in.
Displacement	3.603 cu. in.	4.43 cu. in.	5.22 cu. in.	6.65 cu. in.
Piston Diameter	1.7470–1.7465	1.9360–1.9355	2.1205–2.1200	2.3720–2.3715
Breaker Point Setting	0.020	0.020	0.020	0.020
Coil Air Gap	0.010	0.010†	0.010	678103 Flywheel 0.016 0.010

C/D Air Gap	'			0.010
Spark Plug	Champion J-11-J Gap 0.025	Champion J-14-J Gap 0.025	Champion J-14-J Autolite A11X or equivalent. Gap 0.025	Champion CJ-14 or equivalent. D-400 Series 0.025
Governed Speed	3300 R.P.M.	3200 R.P.M.	3200 R.P.M.	D-600 Series 0.035
Fuel Mixture	½ pint S.A.E. 40 wt. service ML-MM-or Service MM	½ pint S.A.E. 40 wt. service ML-MM-or Service MM	½ pint S.A.E. 40 wt. service ML-MM-or Service MM	3200 R.P.M. ½ pint S.A.E. 40 wt. service ML-MM-or Service MM

Clearances

CRANKSHAFT				
Top Journal	0.6692–0.6689	0.8742–0.8737	0.8742–0.8737	0.8742–0.8737
Crank Pin	0.6865–0.6860	0.7500–0.7495	0.7500–0.7495	0.7430–0.7425
		**0.7430–0.7425	*0.7430–0.7425	
Bottom Journal	0.6692–0.6689	0.8742–0.8737	0.8742–0.8737	0.8742–0.8737
CONNECTING ROD				
Wrist Pin Hole	0.3659–0.3654	0.4275–0.4280	0.4908–0.4903	0.4908–0.4903
Crank Pin Hole	0.6880–0.6875	0.7530–0.7525	0.7531–0.7525	0.9427–0.9422
		**0.9427–0.9422	*0.9427–0.9422	
WRIST PIN				
Diameter	0.3650–0.3642	0.4272–0.4270	0.4900–0.4898	0.4900–0.4898
PISTON				
Diameter	1.7470–1.7465	1.9360–1.9355	x2.1205–2.1200	2.3720–2.3715
Wrist Pin Hole	0.3655–0.3650	0.4275–0.4272	x0.4900–0.4905	0.4900–0.4905
PISTON RINGS				
Diameter	1.740 ± 0.000	1.9375 ± 0.000	2.125±0.000	2.378–2.377
End Gap (In Cyl.)	0.020 ± 0.005	0.020 ± 0.005	0.020±0.005	0.005–0.015
Thickness	0.0935–0.0930	0.093 ± 0.0005	0.098–0.088	0.0615–0.0625
CYLINDER				
Inside Diameter	1.751–1.750	1.940–1.941	xx2.126–2.125	
Sub-Base Bushing	0.683–0.680	——		xx2.378–2.377
SUB-BASE HOUSING				
Sub-Base Bushing	0.683–0.680	——	——	——
CRANKCASE				
Top Bearing	——	0.8770–0.8762	0.08770–0.8762	0.8778–0.8762
Bottom Bearing	——	0.8805–0.8780	0.8805–0.8780	0.8805–0.8780
ARMATURE PLATE				
Bearing	——	0.8770–0.8762	0.8770–0.8762	0.8762–0.8746

*C-16, C-17, and C-42, C-51, C-61, C-81 engines
**C-73 engine
†C-71 engine

xModel C-77 2.188 piston diameter
0.4900–0.4905 wrist pin hole
Model C-78 2.438 piston diameter
0.4900–0.4905 wrist pin hole
xx1967 "D" series and model C-78
2.380–2.381 cylinder diameter

Torque Specifications

Flywheel nut 335/400
Cylinder to crankcase—105/115 in. lbs
Connecting rod—58/70

5 · Clinton

Engine Identification

Clinton engines are identified by a name plate that is installed on the engine at the factory. The name plate contains the serial number and the model number, both of which are necessary when obtaining replacement parts.

Two numbering systems are used to identify Clinton engines. The first one applies to engines made prior to 1961, the second pertains to engines made after that year.

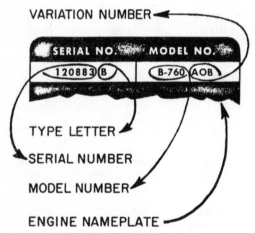

Prior to 1961 engine numbering system

The early numbering system has no practical use to the consumer. It is useful only when ordering parts and only to one who has access to a parts manual. No other information can be gained from the model or serial number than what parts fit a particular engine.

The recent numbering system, however, is quite useful in gaining additional information about a certain engine. The serial number, followed by the type letter, is a numerically sequential number and is used to identify the engine in relation to changes

that are made in design. The model number consists of a ten digit sequence of numbers that is deciphered in the following manner:

The first digit indicates whether the engine is a two stroke or four stroke design. The number 4 indicates a four stroke engine and the number 5 indicates a two stroke engine.

The second and third digits identify the basic engine series and whether the engine has a vertical or horizontal crankshaft. Odd numbers in the third digit position indicate that the engine has a vertical crankshaft and even numbers indicate a horizontal crankshaft.

The fourth digit identifies the type of starter installed on the engine: 0—a recoil starter, 1—a rope starter, 2—an impulse starter, 3—a crank starter, 4—a 12 volt electric starter, 5—a 12 volt starter generator, 6—a 110 volt electric starter, 7—a 12 volt generator, 8—not assigned any specific meaning, 9—indicates a short block.

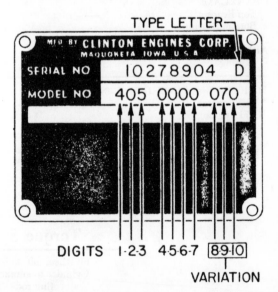

Numbering system after 1961

124

The fifth digit indicates what type of bearing is used on the crankshaft; 1—aluminum or bronze sleeve bearing with a flange mounting surface and pilot diameter on the engine mounting face for mounting equipment concentric to the crankshaft center line; 2—ball or roller bearing; 3—ball or roller bearing with a flange mounting surface and pilot diameter on the engine mounting face for mounting equipment concentric to the crankshaft center line; 4 thru 9—not assigned any meaning.

The sixth digit identifies whether there are any reduction gears or power-take-off units: 0—not equipped with any such unit; 1—auxiliary PTO; 2—2:1 reduction gears; 3—not assigned; 4—4:1 reduction gears; 5—not assigned; 6—6:1 reduction gears; 7 thru 9—not assigned.

The eighth through tenth digits are used to identify model variations.

The type letter identifies parts that are not interchangeable.

Be sure to give the model number and the type letter when obtaining replacement parts.

Carburetion

Clinton engines use a variety of float type carburetors and one type of suction lift carburetor.

A carburetor can only malfunction as a result of these causes: the presence of foreign matter (dirt, water); out of adjustment (too rich or too lean a mixture), and leakage caused by worn parts or cracks in the casting. Look for any of the above causes before disassembling the carburetor. Of course, it may be required that the carburetor be disassembled to correct the problem, but at least you will have some direction.

NOTE: *Check the throttle shaft/main casting tolerance. This is one place on the carburetor that cannot be repaired.*

501 ENGINE CARBURETOR

The 501 engine carburetor is a relatively simple carburetor and can be disassembled after it is removed from the engine.

1. Remove the choke and air filter assembly, remove the mixture adjusting screw

and spring, and unscrew the large bolt which holds the float bowl to the rest of the carburetor.

2. Please note that the main fuel nozzle is contained in the top of the large bolt and care should be exercised not to damage the needle seat. Disassemble the float and fuel inlet needle.

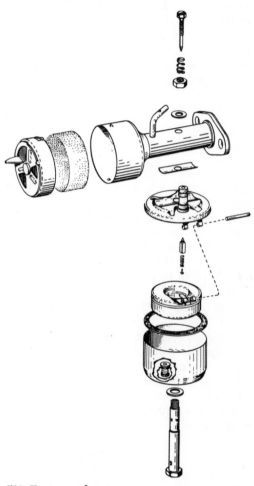

501 Engine carburetor

NOTE: *The needle seat is not replaceable. The bowl cover, needle pin, spring, and seat assembly must be replaced if any part is worn.*

3. Clean all parts in solvent, blow dry with compressed air, inspect, and replace any worn or damaged parts.

4. Assemble the carburetor in the reverse order of disassembly. Set the float level with the bowl cover inverted and the float and needle installed so that there is $1\frac{3}{64}$ in. plus or minus $\frac{1}{32}$ in. clearance between the outer edge of the bowl cover and the free

end of the float. Adjust the level by bending the lip of the float with a screwdriver. *Use all new gaskets.*

5. Set the mixture adjusting needle one turn open from the seat to start the engine. This carburetor does not have an idle mixture orifice and therefore will not operate at speeds below 3000 engine rpm. The operating range is between 3000 and 3800 rpm. Make the final mixture adjustment with a tachometer if possible. If a tachometer is not available, adjust the mixture screw $\frac{1}{16}$ of a turn at a time until you obtain the best engine performance.

SUCTION LIFT CARBURETOR

Rebuilding

THROTTLE PLATE AND SHAFT REPLACEMENT

1. Drill through the plug at the rear of the carburetor body.

2. Force out the plug with a small drift pin.

3. Remove the plastic plug.

4. Remove the screws which retain the throttle plate to the shaft and remove the plate and shaft.

5. Reinstall in the reverse order of removal.

Make sure that the throttle plate and shaft operate without binding. Use a sealer on the new expansion plug.

NOTE: *Horizontal suction lift carburetors do not have a plug.*

The choke plate and shaft is replaced in the same manner as the throttle, except

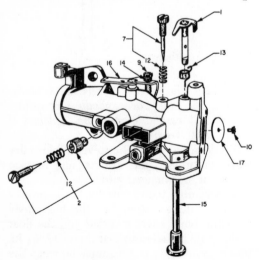

Suction lift carburetor

that there is no expansion plug to be removed.

STAND PIPE REPLACEMENT

The stand pipe is press-fit into the body of the carburetor. Remove the pipe by clamping it in a vise and twisting the carburetor body while at the same time pulling until the pipe is free. To install the pipe, tap it lightly into place until the end of the pipe is 1.94 in. $\pm$ 0.45 in. from the body of the carburetor. Use a sealer at the point where the pipe is inserted into the carburetor body.

IDLE NEEDLE AND JET REPLACEMENT

The idle needle and jet are replaceable and can be removed as follows:

1. Unscrew the needle and remove it.

2. Remove the expansion plug in the bottom of the carburetor body.

3. To remove the idle jet, push a piece of $\frac{1}{16}$ in. diameter rod through the fuel well and up the idle passage until the jet is pushed out of the passage.

To install the idle needle and jet:

4. Make a mark exactly $1\frac{1}{4}$ in. from the end of the $\frac{1}{16}$ in. diameter rod and push the new jet into the idle passage until the mark is exactly in the center of the main fuel well.

5. Install the needle and set it at 4 to $4\frac{1}{4}$ turns open from being seated.

The high speed needle and seat can be inspected for wear or damage by removing the needle and checking its taper. If the needle taper is damaged, replace it. Inspect the seat for taper and splits. If the seat is split or tapered from the needle being screwed in too tight, the complete carburetor must be replaced, since the seat cannot be replaced separately. When the needle is replaced, the initial adjustment is $\frac{3}{4}$ to $1\frac{1}{2}$ turns open from being *lightly* seated.

LMG, LMB AND LMV TYPE CARBURETORS

These carburetors are float type carburetors and can be reconditioned in the same manner as the 501 carburetors by taking note of the following differences in design and specifications.

The main fuel nozzle needle is inserted from the bottom and pushed up through the middle of the float bowl.

The float, fuel inlet needle, and seat are

basically the same as that of the 501 carburetor.

NOTE: *Do not remove the main fuel nozzle from the carburetor body unless it is*

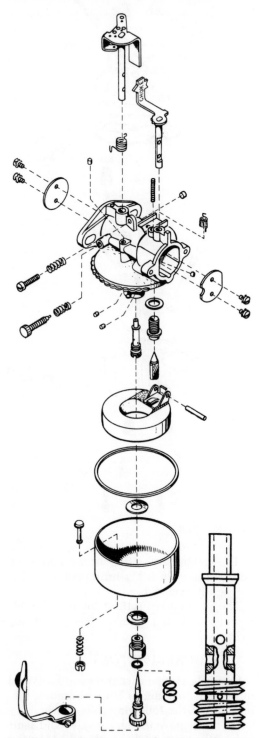

LMG, LMV, LMB type carburetors with an enlarged view of the main jet nozzle

to be replaced. Once it is removed it cannot be reinstalled.

Use all new gaskets.

The throttle valve is installed with the part number or trademark 'W' facing toward the mounting flange.

The preliminary setting of the idle adjusting needle is $1\frac{1}{4}$ turns open after being *lightly* seated.

Adjust the float in the same manner as the 501 carburetor. There should be $\frac{5}{32}$ in. clearance between the float and the casting rim when the carburetor is inverted. When the carburetor is turned over, the float should not drop more than $\frac{3}{16}$ in.

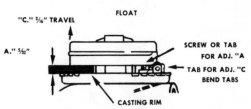

Float adjustment for the LMG, LMV, LMB type carburetors

The preliminary setting for the main fuel adjusting needle is $1\frac{1}{4}$ turns open after being *lightly* seated.

H.E.W. CARBURETORS

H.E.W. carburetors are float type carburetors and can be serviced in the same manner as 501 carburetors. After cleaning and inspecting the parts for wear or damage, assemble the carburetor, noting the following differences:

Install the throttle valve with the part number or 'W' toward the mounting flange with the throttle in the closed position. Always use new gaskets during assembly.

The initial setting for the idle adjusting screw is $1\frac{1}{2}$ turns open from being *lightly* seated.

If any part of the float assembly has to be replaced, replace the whole assembly. Do not replace individual parts.

The float setting is made in the same manner as the 501 carburetor, by turning the main body of the carburetor upside down and measuring the clearance between the float and the body casting rim. The distance in this case is $\frac{3}{16}$ in. with $\frac{3}{16}$ in. of travel.

The preliminary setting for the high speed adjusting screw is $1\frac{1}{2}$ turns open from being seated lightly.

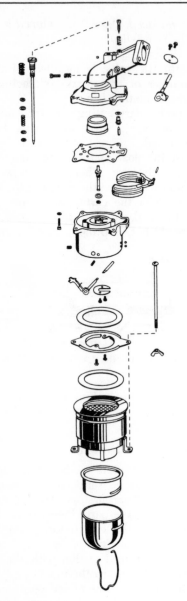

H.E.W. carburetor

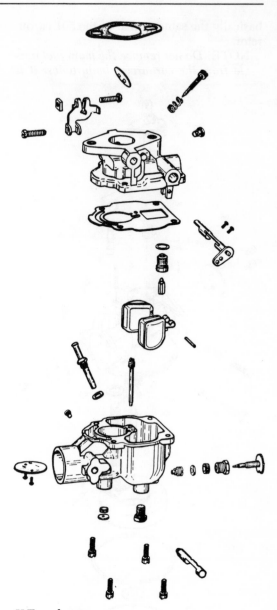

U.T. carburetor

U.T. CARBURETORS

These are also float type carburetors and are serviced in the same manner as 501 carburetors, noting the following differences:

The throttle valve is installed with the trademark 'C' on the side toward the idle port when viewed from the mounting flange side. Use new screws.

Set the float level with the float installed and the housing inverted. The measurement is to be taken with the needle seated and the gasket removed. Measure between the float seam and the throttle body. Adjust by bending the lip of the float.

The preliminary setting for the high speed adjusting needle is $1\frac{1}{4}$ turns open from being lightly seated.

The preliminary setting for the idle adjusting needle is $1\frac{1}{2}$ turns open from being lightly seated.

CARTER CARBURETORS

Carter float type carburetors are serviced in the same manner as 501 carburetors, taking note of the following differences:

Do not remove the choke valve and shaft unless they are to be replaced. A spring loaded ball holds the choke in the wide

Float Level Chart

Carburetor Number	Float Setting (in.)
2712-S	$1\frac{9}{64}$
2713-S	$1\frac{9}{64}$
2714-S	$\frac{1}{4}$
*2398-S	$\frac{1}{4}$
2336-S	$\frac{1}{4}$
2336-SA	$\frac{1}{4}$
2337-S	$\frac{1}{4}$
2337-SA	$\frac{1}{4}$
2230-S	$1\frac{7}{64}$
2217-S	$1\frac{1}{64}$

* When resilient seat is used, set float level at $\frac{9}{32} \pm \frac{1}{64}$.

open position. Be sure to use a new ball and spring when replacing the choke shaft and plate assembly.

Install the throttle plate with the trademark 'C' on the side toward the idle port when viewed from the mounting flange side.

The float setting is made in the same manner as the 501 carburetor and the measurement is $\frac{3}{16}$ in.

The initial setting for the high speed adjusting screw is 2 turns open from being lightly seated.

The initial adjustment for the idle adjusting screw is $1\frac{1}{2}$ turns open from being lightly seated.

Ignition System

Clinton engines use a flywheel type magneto with the coil and breaker point assembly mounted inside the flywheel.

Remove the flywheel to gain access to all of the ignition components, excluding the spark plug, of course.

FLYWHEEL

After removing the flywheel, examine the taper of the crankshaft and the keyways in the crankshaft and flywheel. No bright marks should be visible on the taper of the crankshaft, which would indicate movement of some part. There should be no distortion of any kind on the keyways, as this would indicate a loose flywheel.

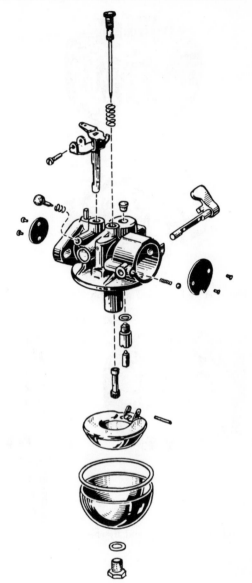

Carter carburetor

Check the magnetism in the flywheel magnets by comparing them to a new part if possible. If the magnets are weak, the flywheel must be replaced.

STATOR

The stator does not normally fail or require service. If the stator has been rubbed by the flywheel or otherwise damaged, replace it.

BREAKER POINTS

1. To check the breaker points, remove the ball, the magneto box cover, and gasket.

2. Look for evidence of excess oil in the

Flywheel and magneto assembly

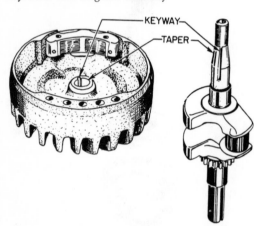

Check the crankshaft and flywheel taper and keyway

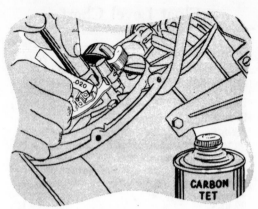

Checking the breaker point gap

box, which would indicate a leaking oil seal or defective breather assembly.

3. Check the point contacts for excessive wear or pitting. Normally the point assembly will last for many years if it is aligned properly, the gap is set correctly, and the rubbing block or shuttle is properly aligned on the actuating cam. If it is necessary to replace the points, make sure that the new set is installed properly.

NOTE: *Do not file the points and only bend the stationary side of the points to align the contact surfaces. Replace the condenser when the points are replaced.*

4. Breaker point gap is critical to engine performance, so make sure it is set to the exact specifications.

BREAKER CAM

1. The breaker cam on most Clinton engines is replaceable. Check the fit of the cam over the crankshaft to make sure that it is tight.

NOTE: *On Model 412 and 413 engines, the breaker cam is machined onto the crankshaft.*

2. Place a little cam grease on the cam and rotate the crankshaft to distribute it evenly after installing a new set of points. Remove any excess grease.

MAGNETO AIR GAP

Magneto air gap is the distance between the stationary laminations of the coil and the rotating magnets of the flywheel. In general, the closer the magnets pass to the laminations, the better the magneto will perform. Some extra clearance must be provided for bearing wear, however. The proper clearance for engines under five horsepower is 0.007 in. to 0.017 in. and 0.012 in. to 0.020 in. for engines over five horsepower.

The air gap is measured by placing layers of plastic tape over the laminations, replacing the flywheel and turning the flywheel. Remove the flywheel and check to see if the flywheel touched the tape, and adjust the coil assembly accordingly. Use only one layer of tape at a time. The thickness of common plastic electrician's tape is about 0.008 to 0.009 in.

Engine Mechanical

The same basic procedure for disassembling a Clinton engine can be used for all engines. The procedure given below pertains to both 2 stroke and 4 stroke Clinton engines; differences are noted. Procedures for servicing individual components are given at a later point in this section.

Engine Disassembly

1. Remove the engine from the piece of equipment it powers and then remove any brackets, braces, adapters or pulleys.
2. Clean the exterior of the engine.
3. Drain the lubricating oil from the crankcase.
4. Remove the fuel tank and blower housing.
5. Remove the carburetor and governor assembly, marking the spring and link holes for correct reassembly.
6. Remove the muffler assembly.
7. Remove the flywheel nut, using a flywheel holder to hold the flywheel while the nut is removed.
8. While lifting up on the flywheel, gently tap the crankshaft to loosen the flywheel from the crankshaft taper. Remove the flywheel and flywheel key.
9. Remove the complete magneto assembly which includes the coil, breaker points, condenser, and laminations.
10. Remove the cylinder head and gasket on 4 stroke engines.
11. Remove the valve chamber cover and breather assembly from 4 stroke engines.
12. Remove the valve spring keepers after compressing the valve spring with a valve spring compressor.
13. Remove the valves and springs from the block after removing the valve spring compressor.
NOTE: *Some valves have a burr on the stem that will prevent the valve from being removed up through the valve guide. If present, it will be necessary to remove this burr from the stem in order to remove the valve from the engine. To remove the burr, hold a flat file against the burred area and rotate the valve.*
14. Remove the base plate or end cover assembly on 4 stroke engines.

NOTE: *On some engines the base is an integral part of the block and cannot be removed. If this is the case, remove the side plate on the power take-off side of the engine. Before removing the side plate or crankshaft, make sure all paint, rust, and dirt are cleaned from the area of the crankshaft bearing. This is so the side plate can be easily removed. When removing the side plate from engines that have ball bearings, it will first be necessary to remove the oil seal and the snap-ring from the crankshaft.*
15. Remove the mounting plate and reed plate assembly on 2 stroke engines.
16. Remove the connecting rod cap screws and cap. Mark the cap and connecting rod so they can be reassembled in the same position.
17. On 4 stroke engines, check for a carbon or metal ridge at the top of the cylinder. If a ridge is present, remove it with a ridge reamer or hone.
18. On 2 stroke engines, push the piston and connecting rod assembly up into the cylinder as far as it will go so it will not hit the crankshaft when it is removed.
19. Remove the piston and connecting rod assembly from a 4 stroke engine's block.
20. Remove the bearing plate on the flywheel side of the block if the engine being disassembled has one.
21. Remove the crankshaft.
NOTE: *On 4 stroke engines with a ball bearing on the PTO side of the engine, it will be necessary to remove the cap screws which hold the bearing in place before the crankshaft can be removed. On some models with ball bearings and tapered roller bearings, it will be necessary to remove the crankshaft oil seal and camshaft axle, and move the camshaft to one side before the crankshaft can be removed. Two stroke engines having a ball bearing on the PTO side of the crankshaft will have to have the retaining ring, which holds the bearing in place, removed before the crankshaft can be removed.*
22. Remove the camshaft assembly. Drive the camshaft axle out of the PTO side of the block as the flywheel side is smaller.
23. Remove the piston and connecting rod assembly from 2 stroke engines.
24. After the camshaft is removed from

a 4 stroke engine, mark the valve tappets as to whether they are the exhaust tappet or the intake tappet, and then remove them. If no valve work is to be done, be sure to replace the valves in the same position from which they were removed.

25. Remove the piston and rod assembly.

After disassembling the engine, clean all parts in a safe solvent, removing all deposits of carbon and oil, etc. Check all operating clearances and replace or rebuild parts as necessary.

CYLINDER BORE

After disassembling the engine, inspect the cylinder bore to see if it can be reused. Look for score marks on the cylinder walls. If there are marks and they are too deep to be removed, the block will have to be discarded. If there is a hole in the block due to connecting rod failure, the block will have to be replaced. If there are broken cooling fins on the outside of the block, these can cause overheating and replacing the block should be considered.

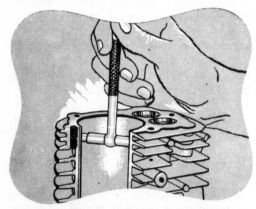

Measuring the cylinder bore—a cutaway view

Check the dimensions of the cylinder bore to determine the extent of wear and whether or not it has to be rebored. The cylinder bore can be rebored to 0.010 in. or 0.020 in. oversize since there are oversize pistons available in these sizes. If the cylinder is within serviceable limits and there is no need to rebore it, be sure to deglaze the cylinder before installing the piston assembly with new rings.

BEARINGS

Clinton engines are equipped with tapered roller bearings, ball bearings, needle bearings, and sleeve bearings. The first thing to determine is whether or not the bearing is worn or damaged and needs replacing. Then clean the bearings in a safe solvent, inspect them for excessive play due to wear, smoothness of rotation, pitted surfaces, and damage. All of these types of bearings are pressed on the crankshaft and into the bearing plates and are removed by either a bearing splitter and puller or they are driven out of the bearing plates with a punch. In all cases be very careful not to bend, gouge, or otherwise damage the crankshaft or bearing plate when removing and installing the bearings.

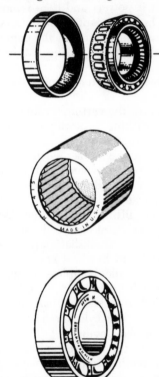

Tapered roller, needle, and ball bearing type bearings, all of which are used in Clinton engines

With sleeve bearings, first inspect the bearing surface for scoring and damage to determine whether the bearing has to be replaced. Check the bearing diameter for wear. In some cases you will find that the bearing surface is the same material as the block. These bearings can be reamed out and sleeve bearings installed. If the original bearing is a sleeve type bearing, it can be removed by driving it out with the proper size driving tool. Install the new bearing so that the oil hole in the bearing is aligned with the oil passage in the block or bearing plate. Drive the bearing into the block or bearing plate until it is recessed about $\frac{1}{32}$ in. from the crankshaft thrust face of the block or bearing plate. After installing the new piece, it must be finish reamed. After finish reaming, clean all metal filings and debris from the engine, making sure that all oil passages are free from obstruction.

VALVE SEATS

Standard valve seats, those without inserts, can be reground to remove all of the oxidized surface metal and gain perfect sealing characteristics. After grinding the valve seats, the valves must be lapped in with lapping compound. Not too much lapping is recommended, just enough to obtain a good seal.

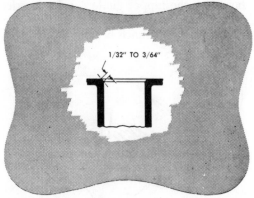

The valve seat width is to be between $\frac{1}{32}$ and $\frac{3}{64}$ in. (0.030–0.045 in.) and the valve seat angle is to be between $43\frac{1}{2}°$ and $44\frac{1}{2}°$.

If the engine has had a number of valve jobs and the valve seat is too deep, requiring that too much stock be removed from the valve stem to obtain the proper valve-to-tappet clearance, valve seat inserts may be installed. If over half of the metal between the lock groove and the end of the valve stem has been removed to gain the proper stem-to-tappet clearance, you should consider installing valve seat inserts.

On aluminum block engines, iron valve seat inserts are standard equipment. To remove these inserts, it is first necessary to remove the metal that has been rolled over the edge of the insert to hold it in place. This is normally accomplished by using the proper size cutter. If the valve seat insert is loose, a cutter may not be necessary. After the insert has been removed, it is necessary to cut the block to the proper depth of $\frac{3}{16}$ in. to $\frac{7}{32}$ in. This is the depth of the insert

plus $\frac{1}{32}$ in. which is used to hold the insert in place. The insert is held to the cylinder block by a definite interference type press fit. The insert should be cooled before attempting to install it in the block. After the insert is fitted in place, with the bevel facing up, the metal around the edge of the insert must be peened over the edge of the insert in order to hold it in place. Do not strike the block too sharply when peening because of the possibility of distorting the cylinder bore. Finish grind the valve seat insert and lap in the valves.

VALVE GUIDES

First, inspect and measure the valve guide diameter to determine whether the guide is worn enough to necessitate rebuilding. The standard guide size for 4 stroke engines under 5 horsepower is 0.2495–0.2510 in. On engines over 5 horsepower, the standard guide size is 0.312–0.313 in.

When the valve stem-to-valve guide clearance is more than the maximum serviceable clearance and cannot be corrected by installing a new valve, you will have to either replace the valve guide (if it is replaceable), oversize it, or knurl it.

Valve guides are replaceable in the 1600, 1800, 2500, 2790, 414-0000-000, 418-0000-000, 420-0000-000, and the 422-0000-000 series engines. The valve guides are pressed out of the block from the base of the cylinder head side. Note the position of the guide before removing it. Install the new guide by reversing the removal procedure. The new guides should be installed at least $1\frac{1}{4}$ in. below the top of the cylinder block. Remove any burrs that might have been created by the installation procedure with a $\frac{5}{16}$ in. (0.312) reamer.

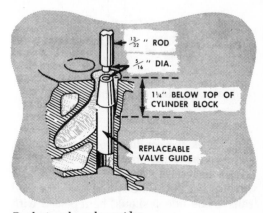

Replacing the valve guides

VALVES

Inspect the valves for a burned face, warped stem or head, scored or damaged stem, worn keeper groove in the stem, and a head margin of less than $\frac{1}{64}$ in. If any one of these conditions exist, the valve should be replaced. Check the valve stem diameter. On engines with less than 5 horsepower, the stem diameter should be 0.2475–0.2465 in. On engines over 5 horsepower, the stem diameter is to be 0.310–0.309 in. Any time the valve stem-to-guide clearance can be reduced more than 0.001 in. by replacing the valve with a new one, you should do so. Also any time the stem-to-guide clearance is over 0.0045 in. you should consider doing some rework (new valve guides, seats and valves) to bring the clearance below 0.0045 in. but not less than 0.002 in. If it is determined that the old valve can be reused, then it should be refaced, using an automotive type valve grinder to secure a 45° face angle on the valve with a $\frac{1}{64}$ in. margin between the head and the face of the valve.

TWO STROKE ENGINE REED VALVES

Inspect the reed valves for the following items: broken reed valves, bent or distorted reed valves, damaged or distorted reed valve seat, or a broken or bent reed valve stop. If any of these conditions exist, the reed valve assembly must be replaced.

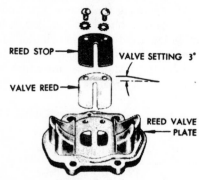

Reed valve assembly for two stroke engines

VALVE SPRINGS

To check the condition of valve springs, simply remove the spring from the engine and stand it on a flat surface next to a new valve spring. If the old spring is shorter and leans to one side, it should be replaced with a new spring. Some of the cast iron engines have a stronger or stiffer valve spring installed on the exhaust valve. Make sure that a stiffer spring is installed on the exhaust valve or, to be sure, install two stiff springs in the engine. When a valve seat is rebuilt, the valve then seats further down into the block and this results in a loss of spring tension. To restore spring tension, install a thin washer on top of the valve spring.

VALVE TAPPETS

The valve tappets should be inspected for wear on the head of the tappet and score marks or burrs anywhere else. The tappet should be replaced if any defects are found. Measure the dimensions of the tappet, checking for stem diameter and length. Oversize tappets are not available; however, the tappet guide can be knurled and rebored to correct size should the tappet-to-guide clearance become too large.

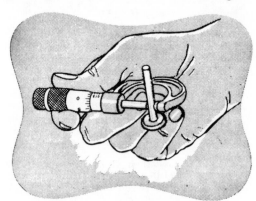

Measuring the tappet stem diameter

PISTON ASSEMBLY

Inspect the piston assembly for scored walls, damaged ring lands, worn or damaged wrist pin lock ring grooves, and a cracked or broken piston skirt. If any of these defects are present, the piston must be replaced. Check the dimensions of the piston with a micrometer.

NOTE: *The ring land diameter on 4 stroke engines is tapered and the reading at the ring land will be 0.00125 in. smaller per 1 in. of piston length.*

Clean all carbon deposits from the ring grooves. An old broken ring will serve as an excellent tool for cleaning ring grooves.

Check the side clearances of the rings with new rings installed. The minimum and maximum clearances for oil, scraper, and

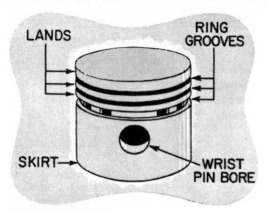

Names of the piston parts

compression rings are as follows: two stroke engines—0.0015–0.004 in.; four stroke engines under five horsepower—0.002–0.005 in.; four stroke engines over five horsepower—0.0025–0.005 in.

If an oversized piston is used, the amount of the oversize is stamped on the top of the piston.

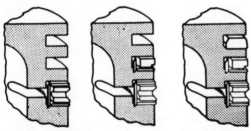

Installation sequence and position for the piston rings

The ring gap on all engines *except* the 1⅞ in. bore two stroke engine is 0.007–0.017 in. The ring gap on the 1⅞ in. bore two stroke engine is 0.005–0.013 in. Oversize rings are available. Install the rings in the following order: oil ring, scraper ring, and compression ring. The oil ring can be

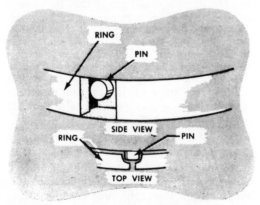

Piston ring retainers in two stroke engines

installed with either side up, the scraper ring should have the step on the lower side toward the bottom of the engine's crankcase and the compression ring has to be installed with the bevel on the inside circumference facing upward.

NOTE: *Two stroke engines have wire retainers or pins located in the ring grooves to keep the ring from moving in the groove. Make sure the ring gap is properly located over these retainers.*

Inspect the connecting rod for wear or damage, such as a scored bearing surface, cracks, and damaged threads. Use a micrometer to check all of the connecting rod dimensions. Check the clearance between the wrist pin and the connecting rod at the wrist pin hole. The tolerance for all Clinton engines is 0.0004–0.0011 in. When the clear-

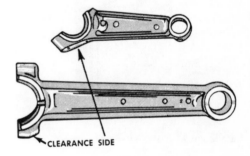

Install clearance rods with the side marked clearance side facing toward the camshaft

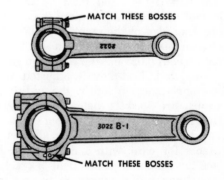

Install the connecting rod cap with the match marks opposite each other

ance reaches 0.002 in., the parts should be replaced with new ones or rebuilt. Clinton, however, does not supply oversize wrist pins.

When installing the piston and connecting rod in a four stroke engine, the piston may go either way, but the rod has an oil hole that must face toward the flywheel side of the engine (with the exception of those engines that use a clearance rod in which the marked side faces toward the camshaft of the engine).

On two stroke engines, the piston is installed with the downward slope of the piston dome facing the exhaust side of the engine. There is no special way for the connecting rod to be installed.

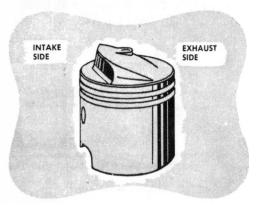

Installation of a two stroke engine piston

CRANKSHAFT

Before removing the crankshaft, remove the spark plug and rotate the crankshaft with the starter mechanism, while checking for any wobble of the end of the crankshaft. Any wobbling indicates that the crankshaft is bent and must be replaced. Deviance of 0.001 in. or more is not tolerable. End-play of the crankshaft should be between 0.008 in. and 0.018 in. If the end-play exceeds 0.025 in., the condition should be corrected. End-play is adjusted when the engine is assembled by the addition of various size gaskets behind the bearing plate. It is not recommended that the crankshaft be straightened. Check all bearing surfaces for wear with a micrometer. Replace the crankshaft if it is bent or cracked; if the keyway is damaged; if the taper is damaged; if the flywheel end threads are stripped; or if the bearings are scored.

CAMSHAFT

Check the camshaft for extremely worn lobes and broken gear teeth. Oil pump drive camshafts have a pin located below the gear that must have a squared end and must be secure to the crankshaft. Camshafts from engines with vertical crankshafts have a scoop riveted to the bottom of the gear. Make sure that the scoop is secure. Make sure that on those models equipped with centrifugal advance (ignition) that the advance mechanism is free and the springs are not distorted or broken. Check the dimensions of the camshaft axle.

CYLINDER

Check the cylinder head for warpage with a straightedge, after removing all dirt and deposits. If the head is warped, place a piece of emery cloth, with the rough side facing up, on a flat surface. Move the cylinder head gasket surface over the emery cloth in a figure eight pattern until the surface of the head is flat. If there are any broken cooling fins or if the spark plug hole threads are stripped, the head must be replaced.

ENGINE ASSEMBLY

Four Stroke Engines

This is a general procedure and is intended to be only a guide since deviations may be necessary for some models.

1. Insert the tappets into the block.
2. Assemble the oil pump to the cam gear, if so equipped.
3. Install the mechanical governor shaft, if so equipped.
4. Install the crankshaft and cam gear into the engine, making sure that the crankshaft thrust washer is in place if one is used.
5. Align the crankshaft and camshaft timing marks.
6. Install the piston and rod assembly using a ring compressor and great caution not to break rings or damage the piston.
7. Install the rod cap and oil dipper, if so equipped, and the cap screw and lock. Tighten to the correct torque. Crimp the screw locks securely.
8. Install the bearing plate and base plate or end cover assembly to the cylinder block. Check the crankshaft end-play. Engines using sleeve bearings should have

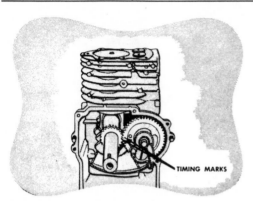

Align the crankshaft and camshaft timing marks

0.005–0.002 in. end-play. Engines using tapered roller bearings should have 0.001–0.006 in. end-play. Engines using roller bearings have no end-play specifications; however, care should be taken not to have the crankshaft too tight after assembly. The end-play is adjusted by the installation of various size gaskets between the plate and block.

9. Install the oil seals in the PTO and flywheel side of the crankshaft.

10. Install the valves into the block and check the valve stem-to-tappet clearance. Clearance is checked with the lobe of the tappet facing away from the valve. Clearance for a four stroke engine is 0.009–0.012 in. Clearance is adjusted by grinding or filing the valve stem.

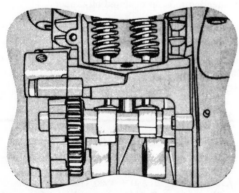

Adjust the valve-to-tappet clearance with the tappets completely off of the cam lobes

11. Using a valve spring compressor, assemble the valve springs to the valves.

12. Install the breather assembly into the valve spring chamber and install the cover.

13. Install the cylinder deflector into the engine.

14. Assemble the magneto assembly to the engine block or bearing plate, which-

ever is applicable, making sure that the points are clean and adjusted to the correct gap.

15. Install the flywheel, flywheel screen, and starter cup to the crankshaft. Tighten the flywheel nut to the proper specification.

16. Install the carburetor assembly and governor, making sure that the governor assembly links and springs are placed in their original holes.

17. Place the cylinder head on the block. Torque the head bolts in three stages, in the proper sequence, and to the proper torque.

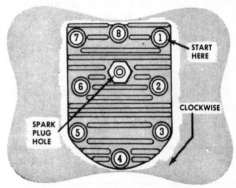

Cylinder head bolt tightening sequence

18. Install the blower housing to the engine.

19. Install the spark plug, muffler assembly, and air cleaner.

Two Stroke Engines

1. Assemble the piston and rod assembly and install it into the block with the help of a piston ring compressor. Be careful not to damage the rings or the piston.

2. Install the crankshaft into the block, installing the crankshaft thrust washer if the engine is so equipped.

3. Assemble the connecting rod and piston assembly to the crankshaft by installing the connecting rod cap and cap screws. Tighten the cap screws to the correct torque.

4. Install the reed valve plate to the engine.

5. Install the bearing plate. Engines that use sleeve bearings should have a crankshaft end-play measuring 0.005–0.020 in. Engines using ball bearings have no specific end-play measurement but make sure that the crankshaft is not tight after assembly. End-play can be adjusted by the addi-

tion of various size gaskets behind the bearing plate.

6. Install the bearing oil seals on the PTO side and the flywheel side of the crankshaft.

7. Install the magneto to the bearing plate and adjust the points to the proper gap.

8. Install the flywheel, flywheel screen, and the starter cup to the crankshaft. Tighten the flywheel attaching nut to the correct torque specification.

9. Assemble the carburetor, governor links, springs, and air vane to the engine. Always replace all governor components in the same position from which they were removed.

10. Install the cylinder deflector to the engine.

11. Install the gas tank.

12. Install the air cleaner, spark plug, and muffler.

Lubrication System

All four stroke engines with vertical crankshafts are lubricated by either a camshaft driven gear type oil pump or an oil scoop attached to the camshaft which rotates in the oil lying in the bottom of the crankcase. The gear type pump forces oil up through a steel tube to the upper main bearing. The oil then falls down onto other parts needing to be lubricated.

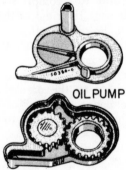
OIL PUMP

Gear type oil pump

The oil scoop attached to the camshaft sprays oil to the top of the engine in a circular path. The upper main bearing in oil scoop engines has an oil access slot in the

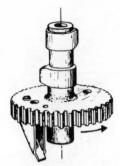
Camshaft mounted oil scoop

top of the bearing area to catch the oil as it is splashed up.

On four stroke engines that have a horizontal crankshaft, there is an oil dipper or distributor located on the bottom of the connecting rod bearing cap. As the crankshaft turns, the dipper churns through the oil in the crankcase and splashes it to all moving parts that need lubrication. There are oil passages, holes, and slots which the oil passes through on its way to the bearing surfaces.

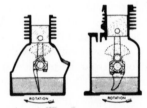
Connecting rod cap mounted oil dipper

All connecting rods have oil access holes or chambers to insure adequate lubrication. Make sure that all oil access holes are installed facing toward the top main bearing in engines with vertical crankshafts.

Recoil Starters

Clinton engines use two basic designs of recoil starters. Service procedures for both starters, including disassembly and assembly, are the same.

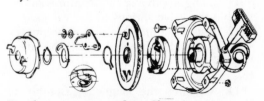
Recoil type starter used on Clinton engines

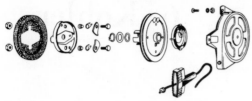

Recoil type starter used on Clinton engines

Assembly and Disassembly

1. Remove the starter assembly from the engine.

2. Slowly release the tension of the recoil spring.

3. Remove the rope pulley assembly.

4. Remove the rope from the pulley.

5. Remove the recoil spring.

6. After disassembling the starter, the parts should be cleaned in a safe solvent and inspected for: bent, damaged, or broken recoil spring; broken, worn, or frayed rope or cable; bent, worn, or broken pawls (dogs) or pawl plate; a bent, cracked, or damaged housing; a worn bearing in the starter rope pulley. If any of these defects are found, the worn or broken part or the whole assembly may have to be replaced.

7. Coat the recoil spring, the inside dome of the housing, and the rope pulley shaft with light grease.

8. Install the recoil spring, making sure that it is installed correctly and that it will wind in the right direction.

9. Install the rope and pulley. Make sure that the recoil spring engages into the pulley. On some models, the rope has to be installed on the pulley before installing the pulley in the starter housing.

10. Assemble the pawls or washers to the starter housing shaft or the rope pulley, whichever is applicable.

11. Wind the recoil spring by turning the rope pulley. The spring is to be wound completely tight and then backed off one turn. Hold the pulley with one hand and install the starter rope and handle if not installed previously. If the recoil spring is wound in the wrong direction or is not backed off one turn, it will break on the first pull of the rope.

6 · Kohler

Engine Identification

An engine identification plate is mounted on the carburetor side of the engine blower housing. The numbers that are important, as far as ordering replacement parts is concerned, are the model, serial, and specification numbers.

The model number indicates the engine model series. It also is a code indicating the cubic inch displacement and the number of cylinders. The model number K181, for instance, indicates the engine is 18 cu in. in displacement and that it has 1 cylinder. The letters following the model number indicate that a variety of other equipment is installed on the engine. The letters and what they mean are as follows:

C   Clutch model
G   Housed with fuel tank
H   Housed less fuel tank
P   Pump model
R   Reduction gear
S   Electric start
T   Retractable start

NOTE: *A model number without a suffix letter indicates a basic rope start version.*

The specification number indicates model variation. It indicates a combination of various groups used to build the engine. It may have a letter preceding it which is sometimes important in determining superseding parts. The first two numbers of the specifications number is the code designating the engine model; the remaining numbers are issued in numerical sequence as each new specification is released, for example, 2899, 28100, 28101, etc. The current specification number model code is as follows:

K91- 26, 27, 28
K141-29
K161-28
K181-30
K241-46
K301-47
K321-60

The serial number lists the order in which the engine was built. If a change takes place to a model or a specification, the serial number is used to indicate the points at which the change takes place. The first letter or number in the serial number indicates what year the engine was built. The letter prefix to the engine serial number was dropped in 1969 and thereafter the prefix is a number. Engines made in 1969 have either the letter "E" or the number "1". The code is as follows:

Letter Prefixes
A—1965
B—1966

ENGINE MODEL	BORE (NOMINAL)	STROKE (NOMINAL)	DISPLACEMENT (CU. IN.)	WEIGHT (APPROX. LBS.)	LUBE OIL CAPACITY (US-QUARTS)	SPARK PLUG GAP	BREAKER POINT GAP
K91	2-3/8"	2"	8.86	41	.5	.025"	.020"
K141 (Spec 29355 & Earlier)	2-7/8"	2-1/2"	16.22	63	1	.025"	.020"
K141 (Spec 29356 & Later)	2-15/16"	2-1/2"	16.9	63	1	.025"	.020"
K161 (Spec 281161 & Earlier)	2-7/8"	2-1/2"	16.22	63	1	.025"	.020"
K161 (Spec 281162 & Later)	2-15/16"	2-1/2"	16.9	63	1	.025"	.020"
K181	2-15/16"	2-3/4"	18.6	63	1	.025"	.020"
K241	3-1/4"	2-7/8"	23.9	116	2	.025"	.020"
K241A	3-1/4"	2-7/8"	23.9	116	*	.025"	.020"
K301	3-3/8"	3-1/4"	29.07	116	2	.025"	.020"
K301A	3-3/8"	3-1/4"	29.07	116	*	.025"	.020"
K321	3-1/2"	3-1/4"	31.27	119	2	.025"	.020"
K321A	3-1/2"	3-1/4"	31.27	119	*	.025"	.020"

Capacity varies from 1 to 1-3/4 quarts.

General engine specifications

C—1967
D—1968
E—1969
First Digit Numbers
1—1969
2—1970
3—1971
4—1972
5—1973

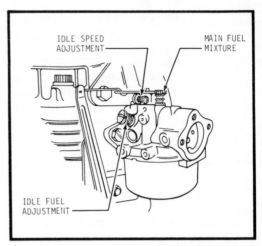

Adjustment screws on the side draft carburetor

Carburetion

Carburetion Adjustments

NOTE: *Before making any adjustments, be sure that the carburetor air cleaner is not clogged. A clogged air cleaner will cause an over-rich mixture, black exhaust smoke, and may lead you to believe that the carburetor is out of adjustment when, in reality, it is not. The carburetor is set at the factory and rarely needs adjustment unless, of course, it has been disassembled or rebuilt.*

1. With the engine stopped, turn the main and idle fuel adjusting screws all the way in until they bottom *lightly*. Do not force the screws or you will damage the needles.

2. For a preliminary setting, turn the main fuel screw out 2 full turns and the idle screw out 1¼ turns.

3. Start the engine and allow it to reach operating temperatures; then operate the engine at full throttle and under a load, if possible.

4. For final adjustment, turn the main fuel adjustment screw in until the engine slows down (lean mixture), then out until it slows down again (rich mixture). Note the positions of the screw at both settings, then set it about halfway between the two positions.

5. Set the idle mixture adjustment screw in the same manner. The idle speed (no-load) on most engines is 1200 rpm; however, on engines with a parasitic load (hydrastatic drives) the engine idle speed may have to be increased to as much as 1700 rpm for best no-load idle.

CARBURETOR REBUILDING

If a carburetor will not respond to mixture screw adjustments, then you can assume that there are dirt, gum, or varnish

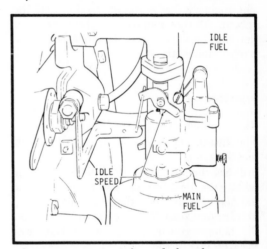

Adjustment screws on the updraft carburetor

deposits in the carburetor or worn/damaged parts. To remedy these problems, the carburetor will have to be completely disassembled, cleaned, and worn parts replaced and reassembled.

Parts should be cleaned with solvent to remove all deposits. Replace worn parts and use all new gaskets. Carburetor rebuilding kits are available.

Disassembly

SIDE DRAFT CARBURETORS

1. Remove the carburetor from the engine.

2. Remove the bowl nut, gasket, and bowl. If the carburetor has a bowl drain, remove the drain spring, spacer and plug, and gasket from inside the bowl.

3. Remove the float pin, float, needle, and needle seat. Check the float for dents, leaks,

and wear on the float lip or in the float pin holes.

4. Remove the bowl ring gasket.

5. Remove the idle fuel adjusting needle, main fuel adjusting needle, and springs.

6. Do not remove the choke and throttle plates or shafts. If these parts are worn, replace the entire carburetor assembly.

Up Draft Carburetors

1. Remove the carburetor from the engine.

2. Remove the bowl cover and the gasket.

3. Remove the float pin, float, needle and needle seat. Check the float pin for wear.

4. Remove the idle fuel adjustment needle, main fuel adjustment needle, and the springs. Do not remove the choke plate or the shaft unless the replacement of these parts is necessary.

Assembly

Side Draft Carburetor

1. Install the needle seat, needle, float, and float pin.

2. Set the float level. With the carburetor casting inverted and the float resting against the needle in its seat, there should be $1\frac{1}{64}$ in. plus or minus $\frac{1}{32}$ in. clearance between the machined surface of the casting and the free end of the float.

3. Adjust the float level by bending the lip of the float with a small screwdriver.

4. Install the new bowl ring gasket, new bowl nut gasket, and bowl nut. Tighten the nut securely.

5. Install the main fuel adjustment needle. Turn it in until the needle seats in the nozzle and then back out two turns.

6. Install the idle fuel adjustment needle. Back it out about $1\frac{1}{4}$ turns after seating it lightly against the jet.

7. Install the carburetor on the engine.

Up Draft Carburetor

1. Install the throttle shaft and plate. The elongated side of the valve must be toward the top.

2. Install the needle seat. A $\frac{5}{16}$ in. socket should be used. Do not over-tighten.

3. Install the needle, float, and float pins.

4. Set the float level. With the bowl cover casting inverted and the float resting lightly against the needle in its seat, there should be $\frac{7}{16}$ in. plus or minus $\frac{1}{32}$ in. clearance between the machined surface casting and the free end of the float.

5. Adjust the float level by bending the lip of the float with a small screwdriver.

6. Install the new carburetor bowl gasket, bowl cover, and bowl cover screws. Tighten the screws securely.

7. Install the main fuel adjustment needle. Turn it in until the screw seats in the nozzle and then back it out 2 turns.

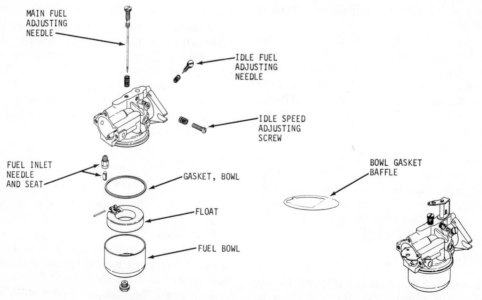

Exploded and assembled view of the side draft carburetor

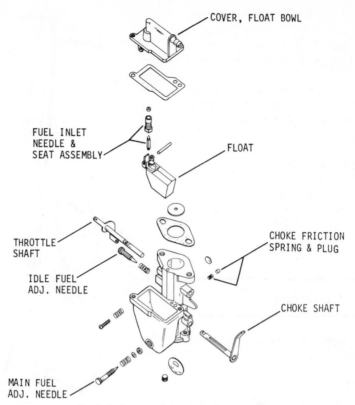

FUEL INLET
NEEDLE &
SEAT ASSEMBLY

COVER, FLOAT BOWL

FLOAT

THROTTLE
SHAFT

IDLE FUEL
ADJ. NEEDLE

CHOKE FRICTION
SPRING & PLUG

CHOKE SHAFT

MAIN FUEL
ADJ. NEEDLE

Exploded view of the updraft carburetor

8. Install the idle fuel adjustment needle. Back it out about 1½ turns after seating the screw lightly against the jet.

Install the idle speed screw and spring. Adjust the idle to the desired speed with the engine running.

9. Install the carburetor on the engine.

Fuel Pump

Fuel pumps used on single cylinder Kohler engines are either the mechanical or vacuum actuated type. The mechanical type is operated by an eccentric on the camshaft and the vacuum type is operated by the pulsating negative pressures in the crankcase. The vacuum type pump is not serviceable and must be replaced when faulty. The mechanical pump is serviceable and rebuilding kits are available.

1. Before disassembling the fuel pump,

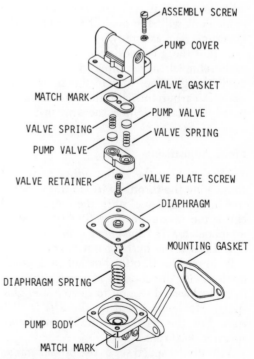

ASSEMBLY SCREW

PUMP COVER

VALVE GASKET

MATCH MARK

PUMP VALVE

VALVE SPRING

VALVE SPRING

PUMP VALVE

VALVE RETAINER

VALVE PLATE SCREW

DIAPHRAGM

MOUNTING GASKET

DIAPHRAGM SPRING

PUMP BODY

MATCH MARK

Exploded view of the fuel pump

mark the upper and lower housing for reference during assembly.

2. Remove the four retaining screws which hold the upper housing to the lower housing.

3. Remove all parts that are to be replaced. To remove the diaphragm, press down and turn the diaphragm 90° to unhook the diaphragm. Install it in a similar manner.

4. Reassemble the fuel pump in the reverse order of disassembly.

NOTE: *It is not necessary to remove the pump from the engine to rebuild it. However, if it is removed, use a new gasket to reinstall.*

Governor

All Kohler engines use mechanical, camshaft driven governors.

Initial Adjustment

1. Loosen, but do not remove, the nut that holds the governor arm to the governor cross shaft.

2. Grasp the end of the cross shaft with a pair of pliers and turn it in counterclockwise as far as it will go. The tab on the cross shaft will stop against the rod on the governor gear assembly.

3. Pull the governor arm away from the carburetor, then retighten the nut which holds the governor arm to the shaft. With updraft carburetors, lift the arm as far as possible, then retighten the arm nut.

Final Adjustment

After making the initial adjustment and connecting the throttle wire on the variable speed applications, start the engine and check the maximum operating speed with a tachometer. If adjustment is necessary:

1. Loosen the bushing nut slightly.

2. Move the throttle bracket in a counterclockwise direction to increase speed, or in a clockwise direction to decrease engine speed. Maximum speed for the K91 is 4000 rpm. For the K141 and K181, maximum speed is 3600 rpm.

3. With the speed set to the proper range,

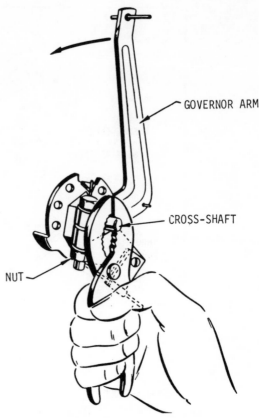

Initial adjustment of the governor installed on the K91, K141, K161, and the K181 engines

tighten the bushing nut to lock the throttle bracket in position.

Ignition System

There are three basic types of ignition systems installed on single cylinder Kohler engines: flywheel magneto types, a battery ignition type, and a breakerless ignition. There are variations on the three basic types in that some are equipped with different capacity charging systems, rotor type and flywheel type magnetos, and starter/generator systems.

SPARK PLUG

The spark plug should be removed, inspected, and replaced if necessary. Set the electrode gap to 0.025 for gasoline engines, 0.018 for gas fuels, and 0.020 in. for shielded plugs. Tighten the spark plug to 18–22 ft lbs.

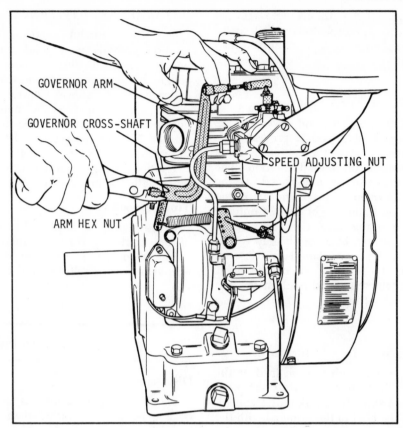

GOVERNOR ARM

GOVERNOR CROSS-SHAFT

SPEED ADJUSTING NUT

ARM HEX NUT

Initial adjustment of the governor installed on the K241, K301, and the K321 engines

BREAKER SYSTEMS

Breaker Point Service

The points are located under the breaker point cover which is under the carburetor. Adjust the point gap to 0.020 in. Replace the condensor when the points are replaced. The light should only glow when connected one way and not the other. If the light glows when connected in both directions, or not at all, the module is defective.

2. Connect one lead of the test light to the "I" terminal and the other to the trigger mounting bracket. If the light comes on, reverse the leads, as the light must be off initially for this test. Lightly tap the magnet with a metal object. The test light should come on and stay on until the leads are disconnected. If the light does not come on, this indicates that the SCR switch is not working properly in which case the trigger module should be replaced.

3. The gap between the projection on the flywheel and the trigger assembly should

be set to 0.005–0.010 in., preferably the closer measurement but no closer than 0.005 in. To adjust the gap, loosen the capscrews on the trigger module and move it in or out as desired.

Ignition Timing

Kohler engines are equipped with a timing sight hole in either the bearing plate or in the blower housing. A snap button may cover the hole on some engines. The button is easily pried loose with a screwdriver so that the timing marks can be observed. Two timing marks are stamped on the flywheel. The "T" mark indicates Top Dead Center (TDC) while the S or SP mark indicates the Spark or Spark Run point which is 20° before top dead center (BTDC).

The same timing procedure is used for the magneto ignition system as for the battery ignition systems. Two methods can be used for timing, the timing light method and the static timing method. The timing light method is more accurate. A timing light can be used with magneto ignition systems but a storage battery will have to be used per the timing light manufacturer's instructions.

Static Timing Method

Remove the breaker point cover and remove the spark plug lead to prevent accidental starting. Rotate the engine by hand in the direction of normal rotation. The points should just begin to break as the S or SP mark or the T mark appears in the center of the timing sight hole. Continue rotating the engine until the points reach maximum opening. Measure the gap with a feeler gauge; the gap should be 0.020 in. fully open. If necessary, loosen the point gap adjustment screw and readjust the gap to the correct setting.

Timing Light Method

1. Remove the high tension lead at the spark plug. Wrap one end of a short piece of fine wire around the spark plug terminal. Reconnect the lead to the terminal with the free end of the wire protruding from under the boot.

2. Connect one timing light lead to the positive or hot side of the battery (whichever terminal is not grounded).

3. Connect the second timing light lead to the wire that is wrapped around the spark plug.

4. Connect the third timing light lead to a ground.

5. Remove the snap button and rotate the engine by hand until the timing mark on the flywheel is visible through the hole. Chalk the timing mark for easy reading.

6. Start the engine and run at 1200–1800 rpm. Aim the timing light into the sight hole. The light should flash just as the timing mark on the flywheel is centered in the sight hole or even with the center mark on the bearing plate or blower housing.

7. If the timing is off, remove the breaker point cover, loosen the gap adjusting screw, and shift the breaker plate until the timing mark is exactly centered. Retighten the adjusting screw before replacing the breaker point cover.

8. Remove the timing light and the wire that was wrapped around the spark plug.

BREAKERLESS IGNITION

The trigger module used on breakerless ignition systems is a solid state device which includes diodes, a resistor, a sensing coil, and a magnet plus an electronic switch called an SCR (Silicon Control Rectifier).

The terminal marked "A" must be connected to the alternator while the terminal marked "I" must be connected to the igniton switch or the ignition coil.

Ignition timing is permanently set on engines with the breakerless system.

Troubleshooting

If the module is suspected as the source of any trouble, perform the following test:

Check for continuity between the two terminals with a self-powered test light.

Engine Mechanical

Disassembly

The following procedure is designed to be a general guide rather than a specific and all inclusive disassembly procedure. The sequence may have to be varied slightly to allow for the removal of special equipment or accessory items such as motor/generators, starters, instrument panels, etc.

1. Disconnect the high tension spark plug lead and remove the spark plug.

2. Close the valve on the fuel sediment bowl and remove the fuel line at the carburetor.

3. Remove the air cleaner from the carburetor intake.

4. Remove the carburetor.

5. Remove the fuel tank. The sediment bowl and brackets remain attached to the fuel tank.

6. Remove the blower housing, cylinder baffle, and head baffle.

7. Remove the rotating screen and the starter pulley.

8. The flywheel is mounted on the tapered portion of the crankcase and is removed with the help of a puller. Do not strike the flywheel with any type of hammer.

9. Remove the breaker point cover, breaker point lead, breaker assembly, and the push-rod that operates the points.

10. Remove the magneto assembly.

11. Remove the valve cover and breather assembly.

12. Remove the cylinder head.

13. Raise the valve springs with a valve spring compressor and remove the valve spring keepers from the valve stems. Remove the valve spring retainers, springs, and valves.

14. Remove the oil pan base and unscrew the connecting rod cap screws. Remove the connecting rod cap and piston assembly from the cylinder block.

NOTE: *It will probably be necessary to use a ridge reamer on the cylinder walls before removing the piston assembly, to avoid breaking the piston rings.*

15. Remove the crankshaft, oil seals and, if necessary, the anti-friction bearings.

NOTE: *It may be necessary to press the crankshaft out of the cylinder block. The bearing plate should be removed first, if this is the case.*

16. Turn the cylinder block upside down and drive the camshaft pin out from the power take-off side of the engine with a small punch. The pin will slide out easily once it is driven free of the cylinder block.

17. Remove the camshaft and the valve tappets.

18. Loosen and remove the governor arm from the governor shaft.

19. Unscrew the governor bushing nut and remove the governor shaft from the inside of the cylinder block.

20. Loosen, but do not remove, the screw located at the lower right of the governor bushing nut until the governor gear is free to slide off of the stub shaft.

ENGINE REBUILDING

Cylinder Block Service

Make sure that all surfaces are free of gasket fragments and sealer materials. The crankshaft bearings are not to be removed unless replacement is necessary. One bearing is pressed into the cylinder block and the other is located in the bearing plate. If there is no evidence of scoring or grooving and the bearings turn easily and quietly it is not necessary to replace them.

The cylinder bore must not be worn, tapered, or out-of-round more than 0.005 in. If it is, the cylinder must be rebored. If the cylinder is very badly scored or damaged it may have to be replaced, since the cylinder can only be rebored to either 0.010 in. or 0.020 in. and 0.030 in. maximum. On the other hand, if the cylinder bore is only slightly damaged, only a light deglazing may be necessary.

Crankshaft Service

Inspect the keyway and the gears that drive the camshaft. If the keyways are badly worn or chipped, the crankshaft should be replaced. If the cam gear teeth are excessively worn or if any are broken, the crankshaft must be replaced.

Check the crankpin for score marks or metal pickup. Slight score marks can be removed with a crocus cloth soaked in oil. If the crankpin is worn more than 0.002 in., the crankshaft is to be either replaced or the crankpin reground to 0.010 in. undersize. If the crankpin is reground to 0.010 in. undersize, a 0.010 in. undersize connecting rod must be used to achieve proper running clearance.

Connecting Rod Service

Check the bearing area for wear, score marks, and excessive running and side clearance. Replace the rod and cap if they are worn beyond the limits allowed.

Piston and Rings Service

PRODUCTION AND SERVICE TYPE

Rings are available in the standard size as well as 0.010 in., 0.020 in., and 0.030 in. oversize sets.

NOTE: *Never reuse old rings.*

The standard size rings are to be used when the cylinder is not worn or out-of-round. Oversize rings are only to be used when the cylinder has been rebored to the corresponding oversize. Service type rings are used only when the cylinder is worn but within the wear and out-of-round limitations; wear limit is 0.005 in. oversize and out-of-round limit is 0.004 in.

The cylinder must be deglazed before replacing the rings. If chrome plated rings are used, the chrome plated ring must be installed in the top groove. Make sure that the ring grooves are free from all carbon deposits. Use a ring expander to install the rings.

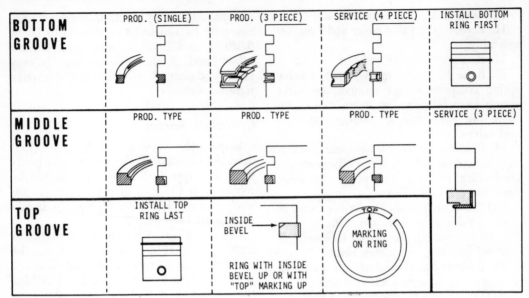

BOTTOM GROOVE	PROD. (SINGLE)	PROD. (3 PIECE)	SERVICE (4 PIECE)	INSTALL BOTTOM RING FIRST
MIDDLE GROOVE	PROD. TYPE	PROD. TYPE	PROD. TYPE	SERVICE (3 PIECE)
TOP GROOVE	INSTALL TOP RING LAST	INSIDE BEVEL → / RING WITH INSIDE BEVEL UP OR WITH "TOP" MARKING UP	TOP ↑ MARKING ON RING	

Positioning of production and service type piston rings on the piston

Piston and Rod Service

Normally very little wear will take place at the piston boss and piston pin. If the original piston and connecting rod can be used after rebuilding, the piston pin may also be used. However if a new piston or connecting rod or both have to be used, a new piston pin must also be installed. Lubricate the pin before installing it with a loose to light interference fit. Use new piston pin retainers whether or not the pin is new.

Valves and Valve Mechanism Service

Inspect the valve mechanism, valves, and valve seats or inserts for evidence of wear, deep pitting, cracks, or distortion. Check the clearance between the valve stems and the valve guides.

Valve guides must be replaced if they are worn beyond the limit allowed. K91 model engines do not use valve guides. To remove valve guides, press the guide down into the valve chamber and carefully break off the protruding end until the guide is completely removed. Be careful not to damage the block when removing the old guides. Use an arbor press to install the new guides. Press the new guides to the depth specified, then use a valve guide reamer to gain the proper inside diameter.

Make sure that replacement valves are the correct type (special hard faced valves

are needed in some cases). Exhaust valves are always hard faced.

Intake valve seats are usually machined into the block, although inserts are used in some engines. Exhaust valve seats are made of special hardened material. The seating surfaces should be held as close to $\frac{1}{32}$ in. in width as possible. Seats more than $\frac{1}{16}$ in. wide must be reground with 45° and 15° cutters to obtain the proper width. Reground or new valves and seats must be lapped in for a proper fit.

After resurfacing valves and seats and lapping them in, check the valve clearance. Hold the valve down on its seat and rotate the camshaft until it has no effect on the tappet, then check the clearance between the end of the valve stem and the tappet. If the clearance is not sufficient (it will always be less after grinding), it will be necessary to grind off the end of the valve stem until the correct clearance is obtained. This is necessary on all engines except the K41, K301, and K321 engines which all have adjustable tappets.

Cylinder Head Service

Remove all carbon deposits and check for pitting from hot spots. Check the cylinder head for flatness. If the head is slightly warped, it can be resurfaced by rubbing it on a piece of sandpaper placed on a flat surface. Be careful not to nick or scratch the head when removing carbon deposits.

Dynamic Balance System Service

The dynamic balance system consists of two balance gears which run on needle bearings. The gears are assembled on two stub shafts that are pressed into special bosses in the crankcase. Snap-rings hold the gears and spacer washers are used to control end-play. The gears are driven off of the crankgear. The dynamic balance system is found on special versions of K241 and K301 models and is standard equipment on K321 engines.

If the stub shafts require replacement, remove the old shafts and press new shafts into the block until they protrude 0.691 in. above the surface of the boss.

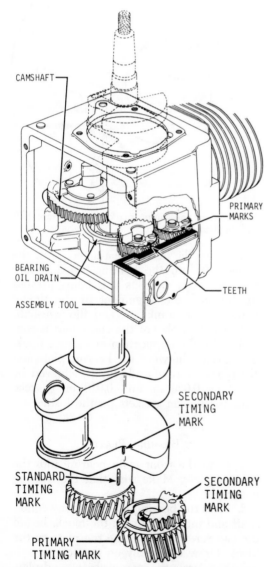

Timing marks for the dynamic balance system

When installing the balance gears, slip one 0.010 in. spacer onto the stub shaft, then install the gear/bearing assembly onto the stub shaft with the timing marks facing out. Proper end-play of 0.005–0.010 in. is attained with one 0.005 in. spacer, one 0.010 in. spacer, and one 0.020 in. spacer which are all installed on the snap-ring retainer end of the shaft. Install the thickest spacer next to the retainer. Check the end-play and adjust it by adding or subtracting 0.005 in. spacers.

To time the balance gears, first press the crankshaft into the block and align the primary timing mark on the top of the balance gear with the standard timing mark next to the crankgear. Press the shaft in until the crankgear is engaged 1/16 in. into the top gear (narrow side). Rotate the crankshaft to align the timing marks on the crankgear and camgear. Press the crankshaft the remainder of the way into the block.

Rotate the crankshaft until it is about 15° past BDC and slip one 0.010 in. spacer over the stub shaft before installing the bottom gear/bearing assembly.

Align the secondary timing mark on this gear with the secondary timing mark on the counterweight of the crankshaft and then install the gear on the shaft. The secondary timing mark will also be aligned with the standard timing mark on the crankshaft after installation. Install the snap-ring retainer, then check and adjust the end-play.

Engine Assembly

REAR MAIN BEARING

Install the rear main bearing by pressing it into the cylinder block with the shielded side toward the inside of the block. If it does not have a shielded side, then either side may face inside.

GOVERNOR SHAFT

1. Place the cylinder block on its side and slide the governor shaft into place from the inside of the block. Place the speed control disc on the governor bushing nut and thread the nut into the block, clamping the throttle bracket into place.

2. There should be a slight end-play in the governor shaft and that can be adjusted by moving the needle bearing in the block.

3. Place a space washer on the stub shaft

and slide the governor gear assembly into place.

4. Tighten the holding screw from outside the cylinder block.

5. Rotate the governor gear assembly to be sure that the holding screw does not contact the weight section of the gear.

CAMSHAFT

1. Turn the cylinder block upside down.

2. The tappets must be installed before the camshaft is installed. Lubricate and install the tappets into the valve guides making sure that the short tappet is installed in the exhaust valve guide on the K141, K161, K181 ACR engines. All other tappets are the same size.

3. Position the camshaft inside the block. NOTE: *Align the marks on the camshaft and the automatic spark advance, if so equipped.*

4. Lubricate the rod and insert it into the bearing plate side of the block. Install one 0.005 in. washer between the end of the camshaft and the block. Push the rod through the camshaft and tap it lightly until the rod just starts to enter the bore at the PTO end of the block. Check the end-play and adjust it with additional washers

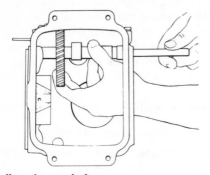

Installing the camshaft

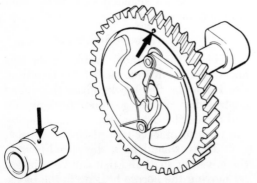

Timing marks for the automatic spark advance

if necessary. Press the rod into its final position.

5. The fit at the bearing plate for the camshaft rod is a light to loose fit to allow oil that might leak past to drain back into the block.

CRANKSHAFT

1. Place the block on the base of an arbor press and carefully insert the tapered end of the crankshaft through the inner race of the anti-friction bearing, or sleeve bearing on the K141.

2. Turn the crankshaft and camshaft until the timing mark on the shoulder of the crankshaft lines up with the mark on the cam gear.

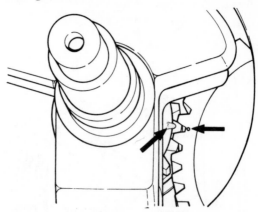

Alignment of the timing marks for the crankshaft and the camshaft

3. When the marks are aligned, press the crankshaft into the bearing, making sure that the gears mesh as it is being pressed in. Recheck the alignment of the timing marks on the crankshaft and the camshaft.

4. The end-play of the crankshaft is controlled by the application of various thickness gaskets between the bearing plate and the block. Normal end-play is achieved by installing 0.020 in. and 0.010 in. gaskets, with the thicker gaskets on the inside.

BEARING PLATE

1. Press the front main bearing into the bearing plate. Make sure that the bearing is straight.

2. Press the bearing plate onto the crankshaft and into position on the block. Install the cap screws and secure the plate to the block. Draw up evenly on the screws.

3. Measure the crankshaft end-play,

which is very critical on gear reduction engines.

PISTON AND ROD ASSEMBLY

1. Lubricate the pin and assemble it to the connecting rod and piston. Install the wrist pin retaining ring. Use new retaining rings.

2. Lubricate the entire assembly, stagger the ring gaps and, using a ring compressor, slide the piston and rod assembly into the cylinder bore with the connecting rod marks on the flywheel side of the engine.

3. Place the block on its end and oil the connecting rod end and the crankpin.

4. Attach the rod cap, lock or lock washers, and the cap screws. Tighten the screws to the correct torque.

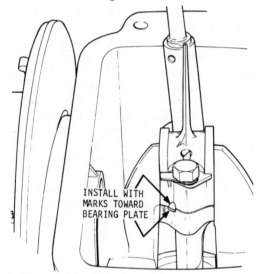

INSTALL WITH MARKS TOWARD BEARING PLATE

Connecting rod and cap alignment marks

NOTE: *Align the marks on the cap and the connecting rod.*

5. Bend the lock tabs to lock the screws.

CRANKSHAFT OIL SEALS

Apply a coat of grease to the lip and guide the oil seals onto the crankshaft.

OIL PAN BASE

Using a new gasket on the base, install pilot studs to align the cylinder block, gasket, and base. Tighten the four attaching screws to the correct torque.

VALVES

1. See the engine rebuilding section of the "Clinton" chapter for details concerning installation of the seats and guides.

2. Place the valve springs, retainers, and rotaters under the valve guides. Install the valves down through the guides, compress the springs, and place the locking keys or pins in the grooves of the valve stems.

CYLINDER HEAD

1. Use a new cylinder head gasket.

2. Lubricate and tighten the head bolts evenly, and in sequence, to the proper torque.

3. Install the spark plug.

BREATHER ASSEMBLY

Assemble the breather assembly, making sure that all parts are clean and the cover is securely tightened to prevent oil leakage.

MAGNETO

Install the magneto coil core assembly and stator on the bearing plate and run the wires through the hole in the bearing plate at the 11 o'clock position.

FLYWHEEL

1. Place the washer in place on the crankshaft and place the flywheel in position. Install the key.

2. Install the starter pulley, lock washer, and retaining nut. Tighten the retaining nut to the specified torque.

BREAKER POINTS

1. Install the pushrod.

2. Position the breaker points and fasten them with the two screws.

3. Place the cover gasket into position and attach the magneto lead.

4. Set the gap and install the cover.

CARBURETOR

Insert a new gasket and assemble the carburetor to the intake port with the two attaching screws.

GOVERNOR ARM AND LINKAGE

1. Insert the carburetor linkage in the throttle arm.

2. Connect the governor arm to the carburetor linkage and slide the governor arm into the governor shaft.

3. Position the governor spring in the speed control disc on the K91, K141, K161, and K181.

4. Before tightening the clamp bolt, turn the shaft counterclockwise with pliers as far

as it will go; pull the arm as far as it will go to the left (away from the carburetor), tighten the nut, and check for freedom of movement.

BLOWER HOUSING AND FUEL TANK

Install the head baffle, cylinder baffle, and the blower housing, in that order. The smaller cap screws are used on the bottom of the crankcase. Install the fuel tank and connect the fuel line.

Starters

REWIND TYPE

Disassembly and Assembly

FAIRBANKS-MORSE

1. Remove the four attaching bolts that hold the starter assembly to the engine and remove it.

2. The starter is disassembled by removing the spring clip in the center and removing the various washers, spring, and then the friction shoes.

3. Release the spring tension by removing the handle and then allowing the rotor to unwind slowly. Prevent the spring from escaping by lifting the rotor about ½ in.

and detaching the spring loop from the rotor.

4. To remove the spring, start with the inside loop and remove the spring carefully from the cover by pulling out one loop at a time, holding back the rest of the turns. Spring holders furnished with the replacement springs simplify the assembly procedure. Place the spring in position in the housing with the outside loop engaged around the pin. Press the spring into the cover cavity thus releasing the spring holder.

5. Assembly of the starter is the reverse of the disassembly procedure.

EATON

1. Remove the five retaining screws that hold the starter assembly to the engine.

2. Remove the screw and washers on the dog retainer; then slip the retainer off of the small spring which is fastened over the post on the outside face of the pulley.

3. To relieve the rewind spring tension, pull the rope handle out about 8 in. and tie a knot in the rope to prevent it from returning. Insert a screwdriver blade under the rope retainer on the handle, slip the rope out of the retainer, and untie the knot at the handle. Hold the pulley sheave with your thumb, to prevent the rewind spring

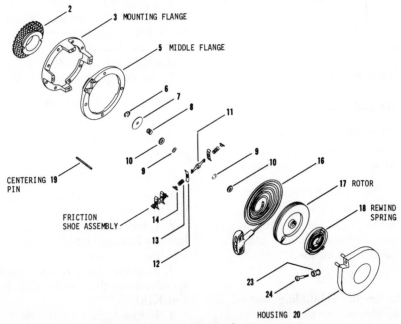

Exploded view of a Fairbanks-Morse starter

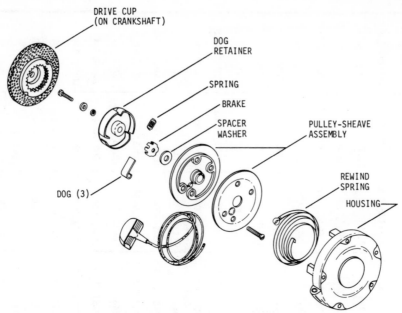

DRIVE CUP
(ON CRANKSHAFT)

DOG
RETAINER

SPRING

BRAKE

SPACER
WASHER

PULLEY-SHEAVE
ASSEMBLY

DOG (3)

REWIND
SPRING

HOUSING

Exploded view of an Eaton starter

from unwinding rapidly, and then untie the other knot and slowly allow the spring to unwind.

CAUTION: *When the pulley and sheave are removed, use extreme caution because the inside loop of the spring fits into the inner hub assembly. The spring can unwind violently unless it is held in the housing while the assembly is removed.*

4. Replace the spring in the same manner as the Fairbanks-Morse starter.

5. To assemble the pulley, sheave, and housing, first bend a piece of wire to form a hook; hook the inside loops of the rewind spring, then pull out to allow the hub on the inside of the pulley to slip into the inside of the spring. Slide the pulley sheave into place with the hub inside the spring. Remove the wire and then fully seat and turn the pulley until the spring engages in the slot on the hub.

6. Assemble the rest of the starter and install it onto the engine in the reverse order of removal.

ENGINE MODEL

SPECIFICATION	K91	K141-161 BEFORE SPEC. # *	K141-161 AFTER SPEC. # *	K181	K241	K301	K321
Bore and stroke	2-3/8x2	2-7/8x2-1/2	2-15/16x2-1/2	2-15/16x2-3/4	3-1/4x2-7/8	3-3/8x3-1/4	3-1/2x3-1/4
Bore diameter, new	2.375	2.875	2.9375	2.9375	3.250	3.375	3.500
Crankshaft end play (free)	.0038/.0228	.002/.023	.002/.023	.002/.023	.003/.020	.003/.020	.003/.020
Crankshaft - conn. rod journal size (Std.)	.9360/.9355	1.1860/1.1855	1.1860/1.1855	1.1860/1.1855	1.5000/1.4995	1.5000/1.4995	1.5000/1.4995
Crankpin - conn. rod side clearance	.005/.016	.005/.016	.005/.016	.005/.016	.007/.016	.007/.016	.007/.016
Crankpin length	.875	1.125	1.125	1.125	1.187	1.180	1.180
Main bearing journal diameter	.984	1.181	1.181	1.181	1.575	1.575	1.575
Connecting rod to crankpin running clearance	.001/.0025	.001/.002	.001/.002	.001/.002	.001/.002	.001/.002	.001/.002
Connecting rod to piston pin clearance	.0005/.0010	.0006/.0011	.0006/.0011	.0006/.0011	.0003/.0008	.0003/.0008	.0003/.0008
Piston pin to piston boss	.0002 Int. to .0002 Loose	.0001 Int. to .0003 Loose	.0001 Int. to .0003 Loose	.0001 Int. to .0003 Loose	.0000/.0003 Select Fit	One Thumb Push Fit	One Thumb Push Fit
Piston to cylinder bore (thrust face)	.003/.004	.0045/.0065	.0045/.0070	.0045/.0070	.003/.004	.003/.004	.0035/.0045
Piston to cylinder bore (top of skirt)	.0035/.0060	.006/.0075	.0060/.0080	.006/.0080	.0075/.0085	.0065/.0095	.007/.010
Piston pin diameter (Std.)	.5624	.6248	.6248	.6248	.8592	.8753	.8753
Ring side clearance, top ring	.002/.004	.0025/.0040	.0025/.0040	.0025/.0040	.002/.004	.002/.004	.002/.004
Ring side clearance, middle ring	.002/.004	.0025/.0040	.0025/.0040	.0025/.0040	.002/.004	.002/.004	.002/.004
Ring side clearance, oil ring	.0015/.0035	.001/.0025	.001/.0025	.001/.0025	.001/.003	.001/.003	.001/.003
Ring end gap	.007/.017	.007/.017	.007/.017	.007/.017	.010/.020	.010/.020	.010/.020
Ring width, top ring	.093	.093	.093	.093	.093	.093	.093
Ring width, middle ring	.093	.093	.093	.093	.093	.093	.077
Ring width, oil ring	.187	.187	.187	.187	.187	.187	.187
Gear reduction shaft end play	.005/.010	.001/.006	.001/.006	.001/.006	.005/.010	.005/.010	.005/.010

Camshaft pin to camshaft clearance	.0010/.0025	.0010/.0035	.0010/.0035	.0010/.0035	.001/.0035	.001/.0035
Camshaft pin to block (Bearing plate end)	.0005/.0012	.0005/.0020	.0005/.0020	.0005/.0020	.0005/.002	.0005/.002
Camshaft pin to block (P.T.O.E.) (Int.)	.0055/.002	.0015/.003	.0015/.003	.0015/.003	.0015/.003	.0015/.003
Camshaft pin to breaker cam	.001/.0025	.0010/.0035	.0010/.0035	.0010/.0035	.0010/.0025	.0010/.0025
Camshaft end play	.005/.020	.005/.010	.005/.010	.005/.010	.005/.010	.005/.010
Valve stem clearance in guide, intake	.0005/.0020	.0010/.0025	.0010/.0025	.0010/.0025	.0010/.0025	.0010/.0025
Valve stem clearance in guide, exhaust	.0020/.0035	.0025/.0040	.0025/.0040	.0025/.0040	.0025/.0040	.0025/.0040
Valve guide in block (Interference)	Not Used	.0005/.0020	.0005/.0020	.0005/.0020	.0005/.0020	.0005/.0020
Valve seat in block (exhaust) (Interference)	.002/.005	.002/.004	.002/.004	.002/.004	.003/.005	.003/.005
Valve clearance, intake (cold)	.005/.009	.006/.008	.006/.008	.006/.008	.008/.010	.008/.010
Valve clearance, exhaust (cold)	.011/.015	.015/.017	.015/.017	.015/.017	.017/.019	.017/.019
Valve seat angle	44.5	44.5	44.5	44.5	44.5	44.5
Valve face angle	45	45	45	45	45	45
Valve seat width	.037/.045	.037/.045	.037/.045	.037/.045	.037/.045	.037/.045
Valve tappet clearance in block	.0005/.002	.0005/.002	.0005/.002	.0005/.002	.0008/.0023	.0008/.0023
Governor bushing to gov. cross shaft clear.	.0012/.0027	.0005/.002	.0005/.002	.001/.0025	.0010/.0025	.0010/.0025
Governor gear to governor shaft	--	.0025/.0055	.0025/.0055	.0005/.0020	.0005/.0020	.0005/.0020
Governor cross shaft end play	--	.005/.030	.005/.030	.005/.030	.005/.030	.005/.030
Ball bearing to cylinder block (Interference)	.007/.0022	.0014/.0029	.0014/.0029	.0006/.0022	.0006/.0022	.0006/.0022
Ball bearing to bearing plate (Interference)	.0007/.0022	.0014/.0029	.0014/.0029	.0012/.0028	.0012/.0028	.0012/.0028
Ball bearing to crankshaft (Int. to loose)	.0004/.0003	.0005/.0002	.0005/.0002	.0004/.0005	.0004/.0005	.0004/.0005
Sleeve bearing to crankshaft (K141 only)	--	.0010/.0025	--	--	--	--
Sleeve bearing to bearing plate, K141 (Int.)	--	.0020/.0045	--	--	--	--

*Specification number for K141 Spec. 28356; for K161 Spec. 28162.

Engine rebuilding specifications

ENGINE MODEL	GOV. ARM LOCK SCREW	CYLINDER HEAD*	CONNECTING ROD*	FLYWHEEL NUT	SPARK PLUG
K91	35 in. lbs.	200 in. lbs.	140 in. lbs.	40-55 ft. lbs.	18-22 ft. lbs.
K141	35 in. lbs.	15-20 ft. lbs.	200 in. lbs.	50-60 ft. lbs.	18-22 ft. lbs.
K161	35 in. lbs.	15-20 ft. lbs.	200 in. lbs.	50-60 ft. lbs.	18-22 ft. lbs.
K181	35 in. lbs.	15-20 ft. lbs.	200 in. lbs.	50-60 ft. lbs.	18-22 ft. lbs.
K241	----	25-30 ft. lbs.	300 in. lbs.	60-70 ft. lbs.**	18-22 ft. lbs.
K301	----	25-30 ft. lbs.	300 in. lbs.	60-70 ft. lbs.**	18-22 ft. lbs.
K321	----	25-30 ft. lbs.	300 in. lbs.	60-70 ft. lbs.	18-22 ft. lbs.

*Lubricate with oil at assembly.
**On earlier K241, K301 Models with 3/4" - 16 thread tighten nut to 100 ft. lbs.

Torque specifications

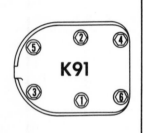
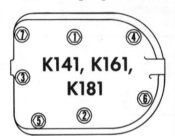
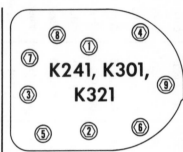

Cylinder head tightening sequences

SPECIFICATION (INCHES)*	K91	K141 & K161 2-7/8" BORE	K141 & K161 2-15/16" BORE	K181	K241	K301	K321
CYLINDER BORE							
Maximum Oversize Diameter	2.378	2.878	2.9405	2.9405	3.2545	3.3785	3.503
Maximum Allowable Taper	.0025	.0025	.0025	.0025	.0015	.0015	.0015
Maximum Out of Round	.005	.005	.005	.005	.005	.005	.005
CRANKSHAFT CRANKPIN							
Maximum Out of Round	.0005	.0005	.0005	.0005	.0005	.0005	.0005
Maximum Taper	.001	.001	.001	.001	.001	.001	.001
CONNECTING ROD							
Maximum Wear Diameter-Big End	.9385	1.1885	1.1885	1.1885	1.5025	1.5025	1.5025
Rod to Crankpin-Max. Clear.	.0035	.0035	.0035	.0035	.0035	.0035	.0035
PISTON - THRUST FACE							
Maximum Wear Diameter	2.359	2.866	2.9305	2.9305	3.2445	3.7025	3.4945
PISTON RING							
Maximum Side Clearance	.006	.006	.006	.006	.006	.006	.006
VALVE STEM TO GUIDE**							
Exhaust - Maximum Clearance	.006	.006	.006	.006	.0065	.0065	.0065
Intake - Maximum Clearance	.004	.0045	.0045	.0045	.0045	.0045	.0045

*Maximum allowable before replacement, reboring, regrinding--see page 15.1-15.2 for new dimensions.
**Measure at top of guide with valve closed.

Maximum wear tolerances and clearances

DIMENSION (SEE FIG. 15-2)	MODEL K91 INTAKE	MODEL K91 EXHAUST	MODEL K141, K161, K181 INTAKE	MODEL K141, K161, K181 EXHAUST	MODEL K241, K301, K321 INTAKE	MODEL K241, K301, K321 EXHAUST
A SEAT ANGLE	89°	89°	89°	89°	89°	89°
B SEAT WIDTH	.037/.045	.037/.045	.037/.045	.037/.045	.037/.045	.037/.045
C INSERT O. D.	------	.972/.973	------	1.2535/1.2545	------	1.2535/1.2545
D GUIDE DEPTH	NONE	NONE	1-5/16	1-5/16	1-15/32	1-15/32
E GUIDE I. D.	NONE	NONE	.312/.313	.312/.313	.312/.313	.312/.313
F VALVE HEAD DIAMETER	.979/.989	.807/.817	1-3/8	1-1/8	1.370/1.380	1.120/1.130
G VALVE FACE ANGLE	45°	45°	45°	45°	45°	45°
H VALVE STEM DIAMETER	.2480/.2485	.2460/.2465	.3105/.3110	.3090/.3095	.3105/.3110	.3090/.3095

Valve specifications

7 · O & R Engines

Engine Identification

There are two numbers to be concerned with when identifying O & R engines: the serial number and the type number.

The serial number is stamped on the mounting flange of the crankcase or on a metal plate attached to the large baffle plate at the cylinder. The serial number is important as it tells when the engine was built. All parts, when they change, are identified by the serial number of the first engine using a new or changed part.

The type number is shown with the serial number and further identifies specific engines that have modifications or adaptations.

Carburetion

Diaphragm carburetors are used on all O & R engines.

Execpt when replacing the carburetor body because of damage, it is generally easier to remove only the diaphragm assembly from the carburetor for service. However, if it is necessary to remove the carburetor, this is done without any other disassembly of the engine.

Diaphragm Removal

1. Remove the carburetor needle valve assembly.
2. Remove the diaphragm screw. This is the long screw closest to the needle valve.
3. Lift off the diaphragm and then take care to remove the diaphragm valve gasket and primer valve.

Carburetor Removal

1. Remove the diaphragm assembly.
NOTE: *The carburetor can be removed with the diaphragm on, but to remove it is faster and eliminates the need to remove and plug the fuel line.*
2. Remove the air cleaner.
3. Remove the lower carburetor screw first. Then remove the top screw which may be difficult to get to. It has to be approached from an angle, so use a good, sharp, #2 bit phillips screwdriver.
4. Remove the gaskets.

Disassembly and Assembly

1. Remove the governor retainer wire. Slide several coils of the spring off the end of the throttle shaft to permit removing the end of the spring from the hole near the end of the throttle shaft.
2. Rotate the spring slightly to disengage the other end of the spring and remove it from the throttle shaft.
3. Remove the two remaining diaphragm screws. Remove the diaphragm cover, gasket, membrane, and disc in that order.
4. Remove the diaphragm arm and spring by lifting up the long end of the diaphragm arm but keeping it engaged under the enclosed end of the diaphragm arm spring. Pull forward and up to remove the spring.
5. Remove the ball valve and diaphragm roller by turning the diaphragm casting over and letting them drop onto a clean cloth.
6. Assemble the carburetor in the reverse order of removal. Replace all worn or damaged parts.

Adjustment

1. Gently seat the needle.
2. Open the needle by turning it counterclockwise ½ turn.
3. Start the engine and run it until it reaches operating temperature.
4. With the engine warm, under load, and the throttle wide open (if so equipped) slowly close the needle until the engine just begins to speed up.

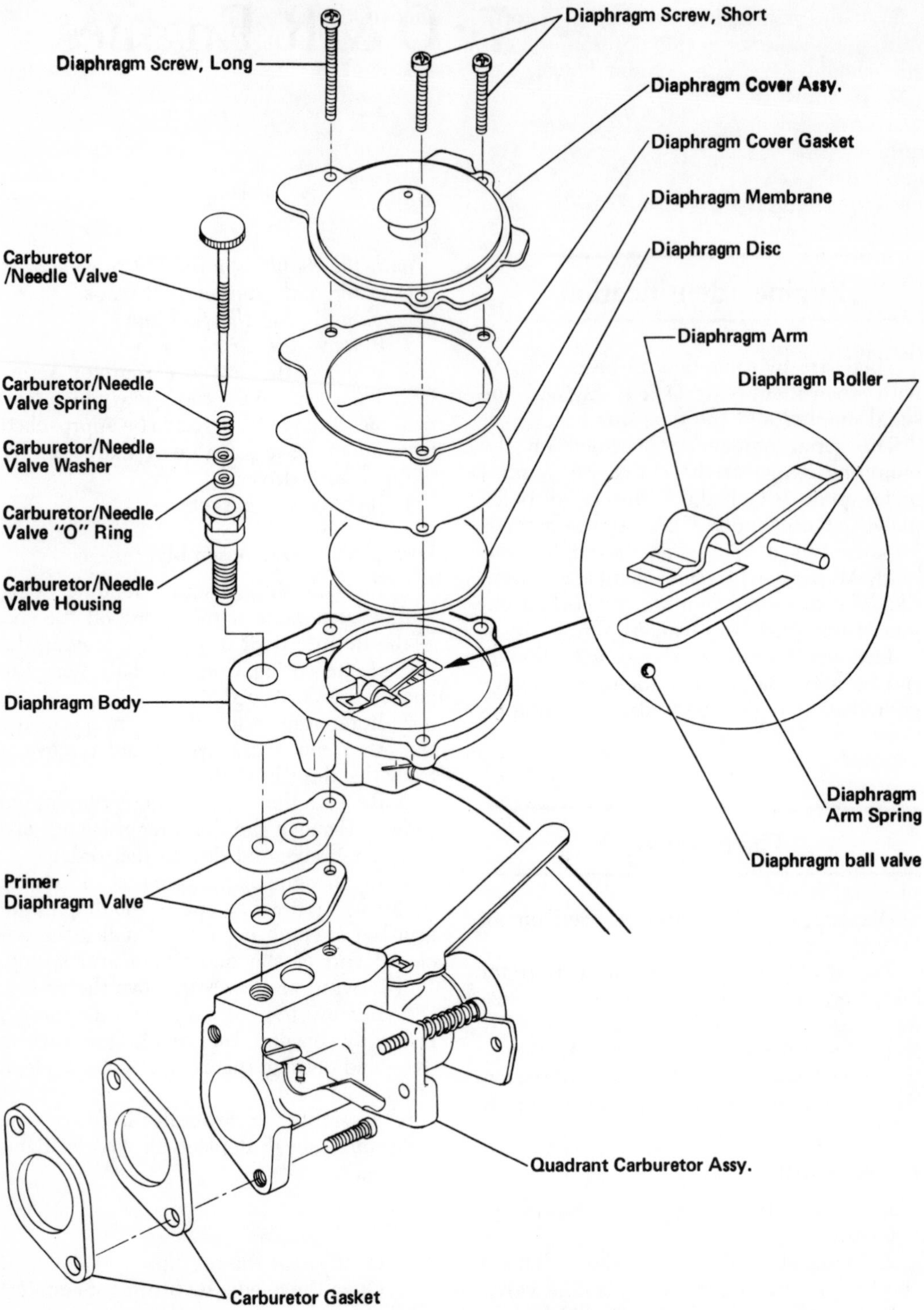

Diaphragm Screw, Long

Diaphragm Screw, Short

Diaphragm Cover Assy.

Diaphragm Cover Gasket

Diaphragm Membrane

Diaphragm Disc

Carburetor /Needle Valve

Carburetor/Needle Valve Spring

Carburetor/Needle Valve Washer

Carburetor/Needle Valve "O" Ring

Carburetor/Needle Valve Housing

Diaphragm Arm

Diaphragm Roller

Diaphragm Body

Primer Diaphragm Valve

Diaphragm Arm Spring

Diaphragm ball valve

Quadrant Carburetor Assy.

Carburetor Gasket

Exploded view of the carburetor

5. At this point, open the needle ⅛ turn. With the needle set in this position, the engine should develop maximum power.

6. To adjust the idle speed, turn the idle speed adjustment screw until the engine is running with the clutch just engaged (on gear drive engines). Back off the idle screw until the clutch just releases. The idle speed should be 2200–2500 rpm.

INDUCTION CASE AND FEATHER VALVE ASSEMBLY

Removal

1. Remove the large cylinder cooling baffle. Note that one of the screws is longer than the others. It must be replaced in the same hole from which it is removed.

NOTE: *If the engine is equipped with a shaft extension and screw, do not remove them until the complete assembly has been removed from the crankcase. If the engine is equipped with the "D" shaft (used with the centrifugal clutch) which does not have the shaft extension and screw, grip the end of the shaft to remove the assembly. To grip otherwise may allow the assembly to come apart causing possible loss of the bearing rollers and greater difficulty in removal of the assembly.*

2. Hold the crankcase section of the engine in one hand and the carburetor/diaphragm/induction case section in the other hand and, with a slight side-to-side motion, remove the assembly. Be careful not to bend or distort the governor vane shaft.

3. Remove the fuel line tube from the diaphragm assembly and plug the line.

NOTE: *While the induction case is removed, it is possible for the connecting rod roller bearing to slide off the crankshaft and into the crankcase. Before replacing the induction case make certain that the bearing is in place.*

4. If the engine is equipped with an extension shaft on the PTO, this must be removed. Loosen the extension shaft screw about 1 turn. Supporting the weight of the induction case assembly by the extension shaft, tap the screw head sharply with a hammer. This will unseat the taper between the extension shaft and the backshaft.

5. Remove the induction case by placing the entire assembly in a vertical position standing on the backshaft counterweight. Gently lift the induction case with a twisting motion.

NOTE: *Watch out for falling roller bearings.*

The induction case sealing ring is replaceable. When a new ring is installed, the open side of the U is toward the small end of the induction case. The feather valve sealing ring can also be replaced, but the open end of the U must go toward the crankcase or large end of the feather valve.

6. Replace the induction case in the reverse order of removal.

Removing the induction case

NOTE: *The feather valve (reed valve) assembly must be replaced if a malfunction is suspected. Do not attempt to repair it.*

Ignition System

Service

To perform any type of service to ignition components, it is necessary to remove the blower housing and the starter as an assembly. Note that engines with a grounding switch on the blower housing have a short wire connecting the coil and the switch. Remove the blower housing carefully and disconnect the grounding wire at the switch.

Engines are built with either clockwise or counterclockwise rotation. Engines with clockwise rotation have the crankshaft keyway for the flywheel directly in line with the connecting rod journal and engines with counterclockwise rotation have the keyway to the left of center in relation with the crankshaft connecting rod journal.

The flywheel retaining nut has a right hand thread for either rotation.

To remove the flywheel, use a spanner wrench to hold the flywheel at either of the two thick lugs in the fins and then loosen the flywheel retaining nut. Turn the nut to the end of the crankshaft and support the engine weight by holding the flywheel. Tap the nut sharply with a hammer to loosen the flywheel from the crankshaft.

NOTE: *The flywheel taper must be dry for installation of the flywheel. Do not use an anti-seize compound.*

Once the flywheel is removed, the ignition breaker points can be serviced. The ignition timing is preset and not adjustable. The correct and accurate setting of the points is very important in this respect. Do not file the points, always replace them. If the breaker points are replaced, the condenser must also be replaced. Set the point gap with the moveable arm of the points on the highest point of the crankshaft breaker cam. Set the gap at 0.015–0.018 in.

Before installing the flywheel, inspect the condition of the keyway and the key. If the keyway or the key is in any way distorted, this would indicate that the flywheel is loose. Install a new key and if the keyway is damaged, a new crankshaft will have to be installed.

Install the flywheel and check the flywheel-to-armature gap. The gap is 0.010 in. Set the gap by turning the flywheel magnets away from the coil core. Loosen the coil core retaining screws, place the correct thickness shim between the coil core and the flywheel, and turn the flywheel with the magnets under the coil core. Secure the coil core position by tightening the retaining screws and then the shim. Turn the flywheel to check the clearance.

Replace the blower housing and the

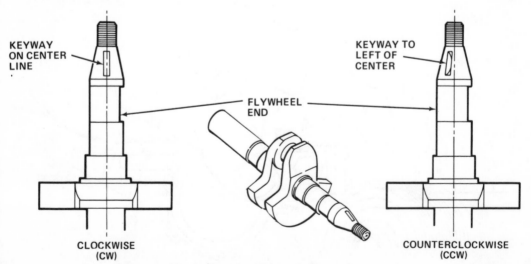

KEYWAY ON CENTER LINE

KEYWAY TO LEFT OF CENTER

FLYWHEEL END

CLOCKWISE (CW)

COUNTERCLOCKWISE (CCW)

Placement of the keyway according to direction of engine rotation

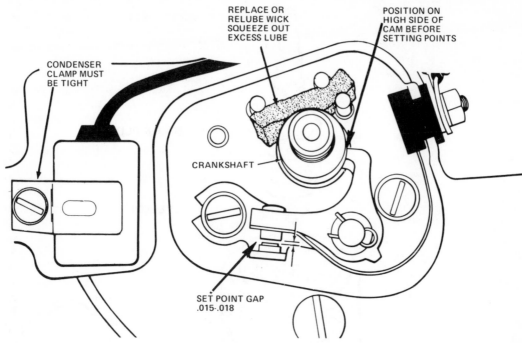

Points and condenser assembly

starter, together with the ignition grounding switch.

Spark plug gap is to be set at 0.025 in.

Engine Mechanical

CYLINDER

Removal

1. Remove the spark plug.
2. Remove the cylinder cooling baffles and the blower housing screws. Remove the blower housing.
3. Insert a cylinder spanner wrench into the slot in the top of the cylinder. Insert a screwdriver through the hole in the wrench and turn the cylinder counterclockwise to thread it out of the crankcase. Pull the cylinder straight up and off of the piston. NOTE: *If the induction case/backshaft assembly has previously been removed from the engine, be careful when handling the crankcase to prevent the connecting rod from sliding off of the crankpin. If the open end of the crankcase is placed down, the connecting rod roller bearing retainer could fall off of the crankpin and the bearing rollers could fall out of the connecting rod.*

4. Turn the cylinder upside down and remove the O-ring cylinder seal, the three heat sink washers, two space washers, the lower exhaust collector gasket, exhaust collector ring, and the upper exhaust collector gasket.

5. Inspect the cylinder for carbon in the exhaust parts and in the top of the cylinder. The carbon should be removed with a screwdriver before reassembling the cylinder to the rest of the engine.

6. Cylinder bore size is 1.250 in. The wear tolerance is 0.0015 in. out-of-round or taper. Lightly hone the cylinder bore to remove the glaze.

7. Remove any carbon from the top of piston and from the ring grooves. Check the rings for wear. A new ring has about 0.007 in. of chrome on the edge of the ring. When the chrome has worn away, the knife edge of the steel ring will be visible as a thin black line around the ring. If the black line is visible, the ring is worn out and should be replaced. Always replace both rings.

NOTE: *Always remove the top ring first and always install the bottom ring first. Turn the rings so that the gaps are*

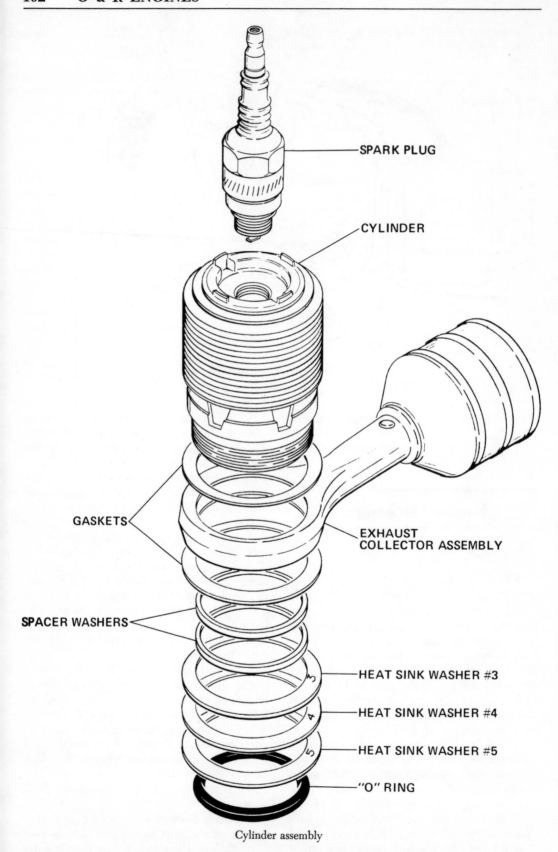

SPARK PLUG

CYLINDER

GASKETS

EXHAUST
COLLECTOR ASSEMBLY

SPACER WASHERS

HEAT SINK WASHER #3

HEAT SINK WASHER #4

HEAT SINK WASHER #5

"O" RING

Cylinder assembly

on opposite sides. The piston, wrist pin, and wrist pin roller bearings cannot be replaced at this time.

Installation

1. Turn the cylinder upside down and install a new upper exhaust collector gasket.

2. Install the exhaust collector ring with the head of the rivet down.

3. Install a new lower exhaust collector gasket.

4. Place the round smooth sides of the two spacer washers together and install them on the cylinder.

5. Install the #3 heat washer next, then the #4 and #5 washer. The numbers are stamped on the top of each heat sink washer. Install the washers with the number facing down.

6. Install the cylinder seal O-ring and roll it all the way down next to the heat sink washer.

7. Move the piston to the top of the stroke and lightly oil the cylinder wall. Slide the cylinder down over the piston.

8. Position the exhaust collector ring/muffler assembly and hold it stationary while threading the cylinder into the crankcase. Do not attempt to do more than start

the cylinder with your fingers. The cooling fins are sharp and can very easily cut your fingers. Use a spanner wrench.

9. Do not force the cylinder to thread onto the crankcase. It can be easily installed if it is properly aligned. Using a cylinder wrench, turn the cylinder until it is absolutely tight or until you can't turn it anymore.

10. Alignment of the cylinder is very important. The slot in the top of the cylinder will be the alignment guide. The slot must be in-line with or at right angles to the crankshaft. When tightening the cylinder, if it is not possible to turn the cylinder another $\frac{1}{8}$ of a turn for proper alignment, it will be necessary to realign it. Do so by turning the cylinder counterclockwise one full turn and then turning it clockwise $\frac{7}{8}$ of a turn to align the slot with the crankshaft.

11. Install the spark plug and the blower housing.

CRANKCASE

Removal

To replace the crankcase, it is necessary to remove all other components of the engine. Proceed as follows:

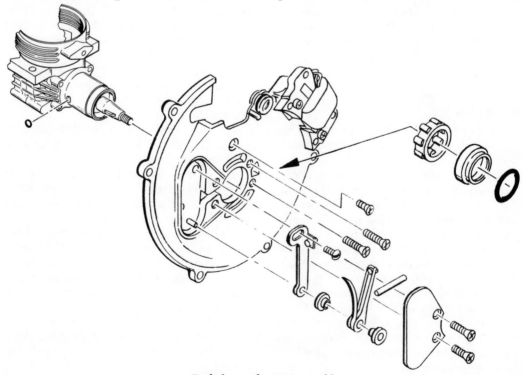

Backplate and points assembly

1. Remove the engine from the machine.

2. Remove the spark plug, the cylinder cooling baffles, the blower housing, the cylinder assembly, the flywheel, and the breaker point/condenser assembly.

3. Remove the four screws which hold the induction case, feather (reed) valve, backshaft, and carburetor assembly to the crankcase. Grip the end of the backshaft and pull the complete assembly away from the crankcase.

4. Remove the connecting rod roller bearing retainer washer.

5. Remove the piston and rod assembly. Be sure to remove all 16 of the roller bearings from the crankcase.

6. Remove the remaining roller bearing retainer washer from the crankpin.

7. Very carefully slide the crankshaft out of the crankcase. The roller bearings will fall out of the bearing retainer. There are 14.

8. Place the crankcase on the bench with the flywheel side of the backplate up. Pry the bearing retainers cup off of the end of the crankcase with a screwdriver.

9. Remove the roller bearing assembly and account for 11 rollers.

10. Remove the 3 remaining screws in the back plate.

11. Before installing a replacement crankcase, make sure that the engine type and serial number are stamped on the flanges of the new crankcase.

Installation

1. Insert a new pushrod O-ring in the recess in the side of the crankcase extension.

2. Install the backplate on the new crankcase. It is possible to press the backplate over the extended end of the crankcase with your hand. However, the backplate must fit flush with the crankcase surface before the screws are installed. Do not attempt to pull the crankcase and backplate together with the screws.

3. Insert the short screw in the single hole near the top of the backplate. Insert two screws in the two holes to the right of the crankcase extension. Do not, at this time, install the two screws which also hold the breaker cover on.

4. Install the large bearing retainer on

the main journal of the crankshaft with the large opening of the retainer facing up.

5. Insert the bearings in the retainer. Use heavy grease to hold the rollers in place. New bearing rollers are covered with wax for this purpose.

6. Install the crankshaft and bearing assembly in the crankcase.

7. Holding the crankshaft in place with one finger, pick up the crankcase and turn it so that the crankcase extension is facing down. Insert the front roller bearing assembly in the bearing race with the closed end of the retainer toward the outside.

8. Continue to hold the crankshaft and turn the crankcase so that the extension and the front bearing are up. Install the front bearing retainer cup and the crankcase sealing ring. The open end of the sealing ring faces down toward the crankcase. Tap the bearing retainer cup with a small hammer to fully seat it on the end of the crankcase.

9. Continue to hold the crankcase in place and install the breaker pushrod (if applicable).

10. Install the breaker points.

11. Install the breaker point box with the two remaining screws (if applicable).

12. Install the flywheel key in the keyway of the backshaft. Turn the shaft so that the key is facing up when the opening in the top of the crankcase for the cylinder is facing up.

13. Line up the key with the flywheel keyway and install the flywheel.

14. Install the washer and nut. Torque the nut to 100 in. lbs.

15. The crankshaft is now located in place. It will no longer be necessary to hold it with your finger.

16. If the connecting rod, piston or piston rings, wrist pin, or wrist pin roller bearings are worn or damaged, they should be replaced at this time.

　a. To remove the piston rings, keep the ring from turning, spread the ring at the gap and lift it out of the ring groove and up over the ring land and the top of the piston. Remove the top ring first and replace the bottom ring first.

　b. Replace the piston by removing the wrist pin retainer lock ring and pushing the wrist pin out the same side the ring was on.

　c. Remove the piston from the con-

NOTE: *The old rollers can be used again if they have not been damaged in anyway. Place a chassis type grease in the openings around the bearing retainer and install the rollers. The grease will keep them from falling out.*

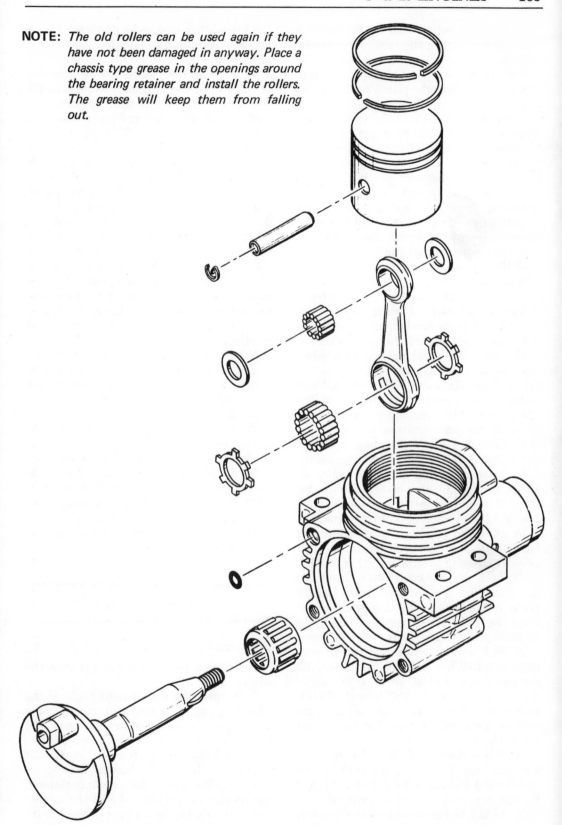

Crankcase assembly

necting rod and account for the 19 roller bearings. Remove the two roller retainer washers.

d. To install a new piston or connecting rod, position the piston upside down. Install a new set of (19) bearing rollers in the wrist pin eye of the rod. Use the wrist pin and a twisting motion to seat the rollers in the eye of the rod. Remove the wrist pin.

e. Position the wrist pin in the piston by pushing it through one side. Push it far enough into the piston to permit the installation of one of the roller bearing retainer washers. Install the washer on the wrist pin so that the flat, sharp side of the washer will be next to the roller bearings.

f. Grease the second roller retainer washer and place it on the eye of the rod opposite the washer on the wrist pin. Put the sharp, flat side next to the rollers. The grease will hold the washer in place while the rod is positioned in the piston.

g. Place the small end of the rod into the piston, align the eye and the rollers with the wrist pin, and push the wrist pin all the way into the piston.

h. Install a new wrist pin lock ring.

i. Install the piston rings if necessary.

17. Place one connecting rod roller retainer washer on the crankpin with the sharp, flat side toward the connecting rod (toward the exposed end of the crankpin).

18. Hold the crankcase assembly in one hand with the flywheel down and the cylinder opening toward the other hand.

19. Turn the crankshaft until the crankpin is at the top of the stroke (in the center of the cylinder opening of the crankcase). Keep it from moving by holding the flywheel with your fingers.

20. Pick up the piston and rod in your other hand. While holding the piston, tilt the piston and rod at about a 45° angle. Insert the eye of the rod into the crankcase so that it will pass easily between the wall of the crankcase and the end of the crankpin.

21. When the eye of the rod is over the crankpin, straighten the angle of the piston and the rod and carefully slide the rod and bearing over and down the crankpin.

22. Install the remaining roller retainer washer with the sharp, flat side next to the rod and rollers.

23. Install a new induction case O-ring into the recess in the upper left-hand corner of the crankcase.

24. Pick up the induction case assembly at the end of the backshaft to prevent the assembly from coming apart. Position the feather (reed) valve casting so that the flat surface of the top will be parallel with the top of the crankcase.

25. Holding the induction case assembly in one hand and the crankcase in the other, position the backshaft slot and the crankpin so that they will match when the induction case is mounted to the crankcase.

26. Push the backshaft onto the crankpin and the induction case up flush against the crankcase. If the induction case will not easily push up to the crankcase, check to see if the feather valve has turned and the flat is no longer parallel to the top of the crankcase. If so, turn it to the correct position.

NOTE: *Do not attempt to pull the induction case flush to the crankcase with the screws. If all of the components are aligned, the induction case will go into place.*

27. Install the long screw through the induction case.

28. Install the three short screws.

29. Install the cylinder assembly to the crankcase. (See "Cylinder Installation.")

30. Install the blower housing and screws, the cylinder cooling baffles and screw, and the spark plug. Install the engine onto the machine.

Recoil Starter

Disassembly

1. Remove the blower housing from the engine.

2. Remove the starter cord knob. If the rope is going to be replaced, cut the rope to remove the knob; otherwise, untie the knot.

3. Hold the end of the cord and let it rewind slowly into the blower housing.

4. Grip the starter reel assembly through the spokes on opposite sides of the ratchet and carefully lift the reel out of the blower housing. Be careful not to disturb the starter spring.

5. Remove the starter cord by unwinding it from the reel and cutting off the anchor wire which holds the end of the cord to the reel.

6. If it is necessary to replace the spring, simply lift it out.

Assembly

1. To replace the starter spring, grip the spring with needle nose pliers about ½ in. to the left of the riveted eye. Holding the spring firmly with the pliers, remove retaining band.

2. Now lower the starter spring into the blower housing and insert the riveted eye of the spring into the slot in the blower housing.

3. Press the spring into the blower housing until all of the coils of the spring rest on the bottom of the cavity. Carefully release the spring from the pliers and allow it to unwind into the full diameter of the cavity.

4. Put the end of the cord through one of the spoke openings in the reel. Pull all of the cord through this opening and let it hang.

5. With the starter reel positioned with the spokes and ratchet up, install the starter reel retainer ring inside the starter reel next to the bottom flange (the side opposite the spokes). Insert the entire reel into the blower housing.

6. Turn the reel slowly until the slot on the starter reel hub engages the hook on the starter spring. When engagement takes place, the reel will drop into the blower housing. If the starter housing fails to engage, check the hook in the end of the starter spring and open it slightly.

7. Using the screwdriver inserted between the ribs of the blower housing, work the starter reel retainer ring down until it is seated in the groove provided in the blower housing.

8. After the retainer ring has been seated, check the opening for the cord in the blower housing. If the ends of the retainer ring are within the cord opening, push the retainer ring around in the blower housing groove until the ends are no longer visible.

9. Placing a screwdriver between the spokes in the starter reel, turn the reel counterclockwise as far as it will go. This will wind the starter spring.

10. Now allow the starter reel to unwind clockwise about one (1) turn.

11. Place your thumb on the top of the starter reel to prevent it from turning and remove the screwdriver.

12. Thread the end of the cord back through the spokes of the starter reel, then through the opening in the blower housing and through the grommet in the casing.

13. Tie a knot in the cord at the starter grommet (older engines have rollers). This will keep the cord from winding back into the starter cord reel.

14. Install the starter knob on the end of the rope.

15. Untie the knot at the grommet and let the cord rewind completely on the reel.

16. Replace the blower housing.

The starter dogs, which are located on the flywheel, are replaced in the following manner:

1. Remove the blower housing.

2. Remove and discard the starter dog retainer washer. This washer must be replaced with a new one.

3. Remove and install the starter dogs one at a time, so that there will always be one in position for reference in replacing the other.

4. After both dogs have been replaced, install the new retainer washer on the flywheel, place a socket over the retainer washer, and tap it with a hammer until it is seated.

5. Replace the blower housing.

8 · Onan

This section pertains to the AJ series of industrial engines and the engines installed on AK series electric generating plants. Although the engines are not the same size in displacement, they are mechanically very similar. Please note that the procedures given here pertain directly to the AJ series industrial engines. They can be applied to the AK generating plant engines with a few very minor changes. Carburetor specifications for both engines are the same.

Setting the governor on the electrical generating plants is more critical than on the non-generator engines because the engine speed determines the power output of the generator.

At the end of the chapter, specifications for size, clearances, and torque for both engine series are given.

Engine Identification

There is an identification plate attached to all Onan engines which contains the model and specification number and the serial number. These numbers are necessary when obtaining parts for the engine. Other information included on the identification plate is the horsepower rating and rpm at which that horsepower is reached.

Carburetion

The carburetor is a side draft or horizontal float type with two adjusting screws for the idle needle and the main fuel nozzle needle.

The carburetor can be disassembled to the extent of removing the air cleaner, removing the adjustment needles, governor

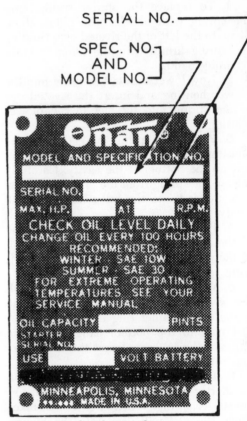

Onan engine identification plate

control linkage, the float bowl, and the float and main fuel inlet needle.

Rebuilding

To rebuild the carburetor, soak all parts in a good quality carburetor solvent or alcohol. Inspect all of the parts for excessive wear or damage, especially the needles, seats, throttle shafts, and shaft bores around which air might leak, and the float hinge. Look for cracks in the housing and warped mounting surfaces. Check the float to make sure that it is not leaking. Reassemble the carburetor using all new gaskets and new replacement parts where necessary. Refer to the "Briggs and Stratton" chapter for specific procedures.

Adjustments

Full-load and no-load adjustments should be made when adjusting the carburetor. After adjustment, it may be necessary to adjust the governor setting, especially on the AK engines which power generators.

To adjust the carburetor, first screw in the adjusting needles gently until they seat. Do not force the needles into their seats as they can be very easily damaged. Back out the main needle about 2½ turns, and the idle needle about ¾ of a turn. Start the engine and allow it to reach normal operating temperatures under a full-load condition.

Turn the main adjustment needle inward until the engine begins to lose speed due to a lean mixture. Then turn the main needle outward until the engine will carry the full load. Check the operation of the engine at various load conditions. If there is a tendency to hunt at any load, turn the adjustment screw outward to obtain a richer mixture, until the problem is corrected. Do not turn the adjusting screw out more than ½ of a turn beyond the point where maximum speed is attained to correct the hunting (an alternate increase and decrease of speed).

To adjust the idle needle, make sure that the engine is warm and running under no-load conditions. Simply turn the needle inward until the engine loses speed due to the lack of fuel, and then back it out until the engine idles smoothly.

The throttle idle stop screw should be adjusted so that the throttle stop lever and the screw clear each other by $\frac{1}{32}$ in. when the engine is operating at the desired speed with no load.

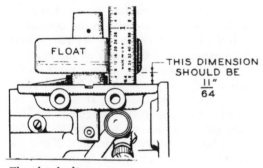

FLOAT

THIS DIMENSION SHOULD BE $\frac{11"}{64}$

Float level adjustment

Ignition Systems

All engines with recoil starters have magneto type ignitions and all engines with electric starters have battery ignitions. For more detailed procedures, refer to the "Briggs and Stratton" chapter.

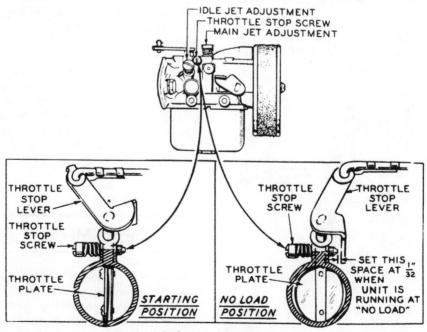

IDLE JET ADJUSTMENT
THROTTLE STOP SCREW
MAIN JET ADJUSTMENT

THROTTLE STOP LEVER

THROTTLE STOP SCREW

THROTTLE PLATE

STARTING POSITION

THROTTLE STOP SCREW

NO LOAD POSITION

THROTTLE PLATE

THROTTLE STOP LEVER

SET THIS SPACE AT $\frac{1"}{32}$ WHEN UNIT IS RUNNING AT "NO LOAD"

Carburetor adjustment

BATTERY IGNITION

The breaker points are located on the side of the engine and are actuated by a plunger. Adjustment of the breaker point gap controls the ignition timing. The gap can be varied from 0.016–0.024 in. to obtain a 19°-25° BTDC spark advance. Decreasing the point gap retards the timing and increasing the point gap advances the timing. Timing marks of 19° and 25° are on the gear cover and a reference mark is on the flywheel. The timing marks can be observed through a hole in the blower housing.

Service

To check for proper operation of the breaker points, turn the crankshaft over until the piston is coming up on its compression stroke. Align the mark on the gear cover with the mark on the flywheel. Adjust the points so that they are just beginning to open. Now turn the flywheel ½ of a turn and check the gap of the points at their greatest separation. The gap should be between 0.016 in. and 0.024 in. If the gap is not between these two specifications, it may be necessary to replace the breaker plunger with a new part.

The points should be inspected for wear, pitting, and misalignment if ignition trouble is encountered. If the points are badly pitted or worn, they should be replaced. If the points are in relatively good shape but are out of alignment (the two contact surfaces do not touch squarely), there is no need to replace the points. They should be aligned by bending the stationary contact. If the contact surfaces have been out of line for some time, there is the possibility that the contacts have become worn out of alignment. If this is the case, replace the point assembly. Every time the contact assembly is replaced, replace the condenser. When the new set of points is installed, make sure that they are clean and free from oil.

MAGNETO IGNITION

Service

To adjust the ignition timing, have the flywheel installed loosely on the crankshaft with the key in place in the keyway. Rotate the flywheel until the mark on the flywheel is in alignment with the proper mark on the gear cover; the points should be just opening. If this is not the case, remove the flywheel and loosen the magneto backing plate mounting screws just enough so that it can be moved with resistance. Move the assembly counterclockwise to advance the timing and clockwise to retard the timing. Tighten the backing plate mounting screws and recheck the gap setting. Replace the flywheel and key and tighten the flywheel nut to the proper torque.

Inspect the contacts of the points to see if they are pitted, worn, or misaligned. Replace the points if the contacts are badly worn or pitted. If the contact surfaces are misaligned, realign them by bending the moveable half of the contact set. Make sure that the contacts are free from oil and dirt.

Ignition breaker assembly for a battery ignition system

Engine Mechanical

Refer to the "Clinton" chapter for a step-by-step disassembly procedure. For detailed instructions and procedures pertaining to engine rebuilding, refer to the "Briggs and Stratton" chapter.

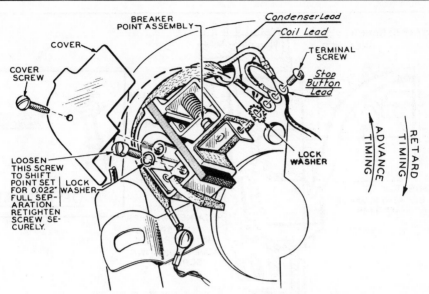

Ignition breaker assembly for a magneto ignition system

VALVES

Service

In order to remove the valves, the cylinder head must be removed. Remove the cylinder head screws and then remove the head from the engine. If the head sticks to the block, hit it lightly with a soft hammer, being careful not to damage any of the cooling fins. Remove the cylinder head gasket and discard it.

Use a conventional type valve spring compressor to compress the valve springs so that the spring retainer can be removed. The retainers are the split, tapered type and will most likely fall out when the valve spring is compressed. After removing the retainers, lift the valve out through the top of the valve guide. Clean the valves of all carbon deposits and inspect them, looking for warpage, worn stems, and burned surfaces that are partially destroyed. Determine whether or not the valve can be reused, whether it can be reground, or if the valve has to be replaced.

Check out the valve stem-to-guide clearance and, if it is too large, the guides must be replaced. They can be removed through the valve chamber. The valve tappets are also replaceable from inside the valve chamber, once the valves have been removed.

The valve face angle is 44°. The valve seat angle is 45°. The 1° interference angle assures a sharp seating surface between the valve and the seat and good sealing characteristics. The valve seat width must be between $\frac{1}{32}$ in. and $\frac{3}{64}$ in. Valves should not be hand lapped if at all possible. This is especially important if stellite valves are used.

To check the valves for a tight seal, make pencil marks around the valve face, then install the valve and rotate it a part of a turn. If the marks are all rubbed off uniformly, then the seal is good.

Lightly oil the valve stems and reassemble all of the valve parts in the reverse order of removal. Adjust the valve tappet clearance.

Adjustment

To adjust the valve tappet clearance, first remove the valve compartment cover and blower housing, if they are not already removed. Crank the engine over until the intake valve, which is the one nearest to the carburetor, opens and closes. Continue to turn the crankshaft slowly until the timing mark on the flywheel is aligned with the TDC mark on the gear cover. This places the valve tappets off of the lobes of the camshaft. Measure the clearance between the valve tappet and the stem of the valve. The specification table is at the end of this chapter. The tappets are fitted with self-locking screws. Use a $\frac{9}{16}$ in. wrench to hold the tappet and a $\frac{7}{16}$ in.

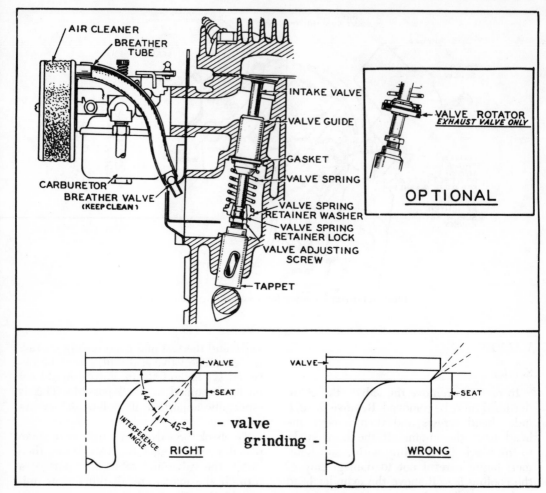

Valve and crankcase breather assembly

wrench to move the adjusting screw when adjusting the valves.

GEAR COVER

Removal and Installation

In order to gain access to the camshaft gear and other internal components of the engine, it is necessary to remove the gear cover. In order to remove the gear cover, the magneto assembly must be removed. Disconnect the spark plug wire at the spark plug and disconnect the stop wire. Remove the attaching screws that hold the magneto assembly to the gear cover and remove the magneto.

When the gear cover is removed, the governor shaft disengages from the governor cup, which is part of the camshaft.

During installation of the cover, be sure to engage the pin on the cover, with the chamfered hole located in the governor cup.

To do this, turn the governor cup so that the hole is located at the top or in the 12 o'clock position. Turn the governor shaft clockwise as far as it will go and hold it there until the cover is installed. Position the cover on the engine and make sure that it fits flush against the engine. Be careful of the gear cover oil seal during installation. Use a new gasket if the old one is damaged.

GOVERNOR CUP

Removal and Installation

The governor cup can be removed from the camshaft gear. To do so, first remove the gear cover. Remove the lock ring from the end of the camshaft center pin. When the cup is removed, the governor fly balls will fall out from behind the cup. Catch them in your hands. The majority of the engines use 10 fly balls. High speed engines use 5 fly balls.

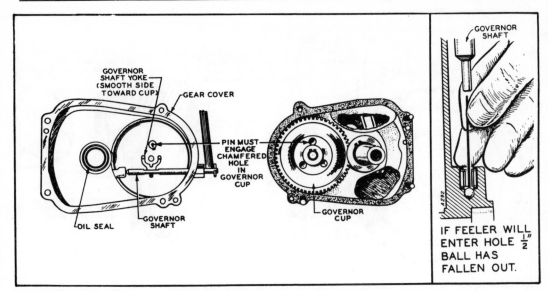

Gear cover assembly

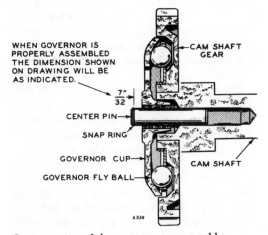

WHEN GOVERNOR IS
PROPERLY ASSEMBLED
THE DIMENSION SHOWN
ON DRAWING WILL BE
AS INDICATED.

Cutaway view of the governor cup assembly

To replace the governor cup and governor fly balls, tip the engine back if possible. Place the balls into position and install the governor cup. When the cup is installed onto the camshaft, there should be a distance of exactly $7/32$ in. between the small lock ring and the center pin of the governor cup. Make certain the cup is pressed back against the fly balls as far as possible. If the distance is too small, file some metal from the face of the cup in order to gain the correct clearance. Remove any and all burrs caused by filing. If the distance is more than $7/32$ in., the center pin can be pressed in by the amount required. Be careful not to damage the pin.

TIMING GEARS

Removal and Installation

If it becomes necessary to replace either the crankshaft or camshaft gears because of broken teeth, extreme wear, cracks, etc., both gears must be replaced as a pair. Never replace only one of the gears. Both gears are pressed onto their respective shafts.

To remove the crankshaft gear, insert two #10-32 screws into the threaded holes in the gear and tighten the screws alternately a little at a time. The screws will press up against the crankshaft shoulder and force the gear off the end of the crankshaft. On AK engines use a gear puller to remove the crankshaft gear.

To remove the camshaft gear, it is necessary to remove the entire camshaft assembly from the engine. First remove the crankshaft gear lock ring and washer. Remove the cylinder head, valve assemblies, fuel pump (if so equipped), and the valve tappets. Remove the governor cup assembly and then remove the camshaft and gear assembly from the engine. The camshaft gear may now be pressed off the camshaft. Do not press on the camshaft center pin as it will be damaged. The governor ball spacer is press fit into the camshaft gear.

When the camshaft gear is replaced on the camshaft, be certain that the gear is properly aligned and the key properly posi-

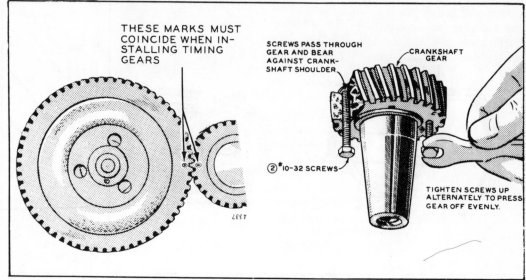

THESE MARKS MUST COINCIDE WHEN IN- STALLING TIMING GEARS

SCREWS PASS THROUGH GEAR AND BEAR AGAINST CRANK- SHAFT SHOULDER

CRANKSHAFT GEAR

2 #10-32 SCREWS

TIGHTEN SCREWS UP ALTERNATELY TO PRESS GEAR OFF EVENLY.

Aligning the timing marks on the timing gears and removing the crankshaft gear

tioned before beginning to press the gear onto the camshaft.

Install the governor cup before replacing the camshaft and gear assembly back into position in the engine.

There are two stamped 'O' marks, one on each gear, near the gear teeth. When the crankshaft and camshaft gears are meshed, these marks must be exactly opposite each other. When installing the camshaft gear assembly, be sure that the thrust washer that goes behind the camshaft gear is installed. Replace the retaining washer and lockwasher on the crankshaft.

CYLINDER BORE, PISTON, AND PISTON RINGS

The cylinder can be rebored if it becomes heavily scored or badly worn. If the cylinder bore becomes cracked, replace the cylinder block.

The cylinder can be bored out to 0.010 in., 0.020 in., or 0.030 in. oversize.

NOTE: *The cylinder bore on the AK engine can be bored out to 0.040 in. oversize.*

There are pistons and rings in the above oversizes available to accommodate an oversized cylinder bore. If the cylinder bore has to be bored out only 0.005 in. to remove damage in the cylinder, use standard size parts.

Use a ridge reamer to remove the ridge that may be present at the top of the cylinder bore to avoid damaging the piston

rings when the piston and connecting rod assembly is removed. Hone the cylinder and create a cross hatch pattern on the cylinder walls if new rings are being installed. Clean the cylinder with SAE 10 engine oil after honing.

Some engines were originally built with 0.005 in. oversize pistons and are so indicated by a letter 'E' following the serial number stamped on the identification plate and on the side of the crankcase.

The piston is fitted with two compression rings and one oil control ring. When the piston assembly is removed from the engine, clean off all carbon deposits and open all of the oil return holes in the lower ring groove. Before installing new rings, check the ring gap by installing the rings squarely in the cylinder bore and measuring the gap between the two ends of the rings. If the gap is too small, it is possible to file the ends to obtain the proper size gap.

Tapered type rings are usually marked with the word 'TOP' on one side. This side must be installed facing toward the top or closed end of the piston. Position the ring gaps evenly around the circumference of the piston with no ring gap over the piston pin.

The piston pin is held in place by two lock rings, one at each end. Make sure that the lock rings are properly installed before installing the piston assembly in the cylinder.

Be sure to check the size of the piston,

piston pin, piston pin bore, and the size of the cylinder before installing any of these parts back into the engine. Replace any parts that are worn beyond the maximum allowed specification.

CONNECTING ROD

Before removing the connecting rod from the crankshaft, mark the cap and rod so they can be installed in exactly the same position from which they are removed.

If abnormal bearing wear (worn on one side more than the other) is noticed, this would indicate that the connecting rod is bent. It is possible to have the connecting rod straightened, but this should be done at a machine shop or small engine service shop.

Measure all of the bearing surfaces for size, including the piston pin hole and the crankpin bearing. It is possible to reduce the size of the clearance between the connecting rod bearing and the crankshaft journal by carefully moving the mating surfaces of the connecting rod cap across a piece of #320 grit or less abrasive cloth that is attached to a flat surface such as a piece of glass.

Be sure to reinstall the oil dipper to the connecting rod cap when assembling the piston and connecting rod assembly to the crankshaft.

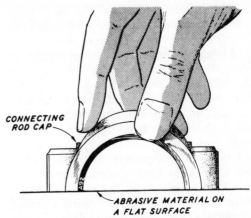

Reducing the clearance between the crankshaft bearing journal and connecting rod by sanding the rod cap

BEARINGS FOR THE AJ SERIES

The main bearings are sleeve bearings which are flanged so as to take the thrust of the crankshaft. On early models with a specification letter 'A', lead babbit faced bearings with no flange were used. These bearings are not interchangeable with the later design bearings.

When replacing the bearings, the main bearings must be pressed into the bearing plate from the inside. The bearings must then be line bored or reamed to the correct size.

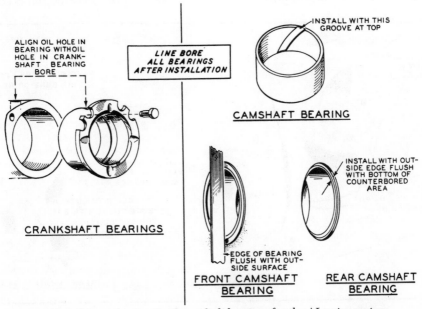

Crankshaft main bearings and camshaft bearings for the AJ series engines

NOTE: *Replacement of the main bearings should not be attempted unless the proper equipment is available.*

When positioning the bearings on the engine, align the oil hole in the bearing with the oil hole in the block bore. In engines that are splash lubricated, the hole will be upward; on engines that are pressure lubricated, the hole will be opposite the camshaft.

Crankshaft end-play is adjusted by placing various thickness gaskets between the rear bearing plate and the cylinder block. When the bearing plate is being installed on engines that are pressure lubricated, the hole in the bearing plate gasket is to be aligned with the oil hole in the bearing plate.

The camshaft bearings are steel backed, babbit lined, and not flanged. The front bearing is pressed into the cylinder block until it is flush with the outside surface of the cylinder block. The rear bearing is pressed in until it is flush with the bottom of the counterbore in which the expansion plug is installed.

BEARINGS FOR THE AK SERIES ENGINES

The main crankshaft bearings installed in AK series engines are precision bearings. They are available in the standard size as well as 0.002 in., 0.010 in., 0.020 in., and 0.030 in. undersize. These precision type bearings are *NOT* to be line reamed.

The bearings are press fit into the cylinder block. Before trying to install the bearing into the cylinder block or bearing plate, heat the plate or block by running hot water over them or placing them in an oven heated to 200° F. This will cause the block or bearing plate to expand and facilitate the installation of the sleeve bearing.

The oil hole in the bearing and the oil hole in the bearing bore must be aligned when the bearings are installed. On pressure lubricated engines, the hole should be opposite the crankshaft. On engines that are splash lubricated, the hole should be upward.

Install the cold precision bearing with the inside of the main bearing $\frac{1}{16}$ in. to $\frac{3}{32}$ in. away from the inside end of the bore to allow clearance for the radius of the end of the crankshaft. The end-play of the crankshaft is adjusted in the same manner as the AJ series engines.

Before mounting the AK generator to the engine, tighten the rear bearing plate nuts. After the generator is secured to the engine, sharply strike the flywheel screw to readjust the forward end-play of the crankshaft. This must be done to prevent the excessive front bearing wear.

CRANKSHAFT OIL SEALS

The crankshaft oil seals are installed with the open sides facing toward the inside of the engine. To replace the rear oil seal, the rear bearing plate must be removed. To re-

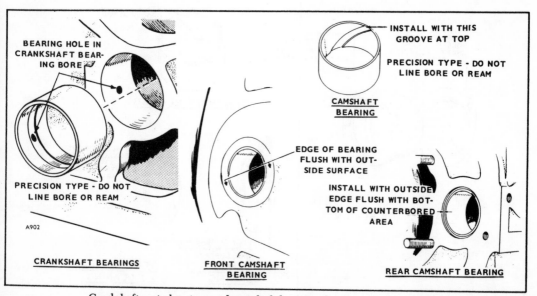

Crankshaft main bearings and camshaft bearings for the AK series engines

place the front oil seal, the front gear cover must be removed. Be careful not to damage the oil seal during installation or to turn back the edge of the seal lip.

Lubrication

Onan engines are either splash lubricated or pressure lubricated.

The splash lubrication system consists of an oil dipper attached to the connecting rod cap and various oil passages to catch and channel the oil that is splashed up by the oil dipper.

The pressure lubrication system consists of a gear type oil pump, an oil intake cup, a non-adjustable pressure relief valve, and the various oil passages and channels.

Pump Removal and Installation

To remove the oil pump, it must be turned off the intake pipe. If the oil pump fails, it must be removed and replaced with a new one. The relief valve can be removed to be cleaned. The oil can be removed and the pressure relief valve replaced if it is damaged.

When installing the oil pump, be sure to use a new mounting gasket. Install the intake pipe so that it is tight and at the right angle to have the cup parallel to the base.

Prime the oil pump with oil before starting the engine. This is very important as the pump will not operate if it is not primed.

Governor

The governor is a mechanical device that uses centrifugal force to control the speed of the engine.

Correct adjustment of the governor mechanism is necessary for satisfactory engine performance. However, adjustment of the governor assembly will not compensate for parts that are worn and causing the governor to malfunction.

When it becomes apparent that the governor needs adjustment, the first thing to do is inspect the governor assembly for binding bearings on the shaft which extends from the gear cover, a binding throttle shaft, and binding at the ball joint. Check for looseness or excessive wear of the governor linkage and mechanism. Excessive play in the linkage will cause the governor to make the engine accelerate and decelerate erratically. This is called hunting. If, after adjusting the governor and making sure that the carburetor is correctly adjusted, the engine still does not perform satisfactorily, replace the governor spring. After a period of long service, a spring will lose its

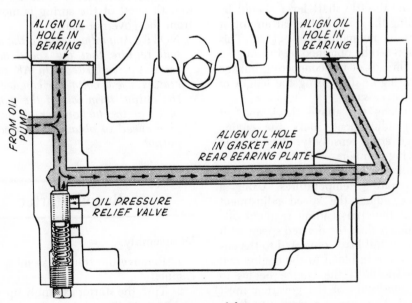

Diagram of the pressure lubrication system

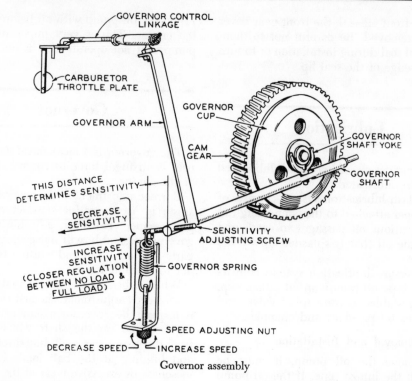

GOVERNOR CONTROL
LINKAGE

CARBURETOR
THROTTLE PLATE

GOVERNOR ARM

GOVERNOR
CUP

CAM
GEAR

GOVERNOR
SHAFT YOKE

GOVERNOR
SHAFT

THIS DISTANCE
DETERMINES SENSITIVITY

DECREASE
SENSITIVITY

INCREASE
SENSITIVITY
(CLOSER REGULATION
BETWEEN NO LOAD &
FULL LOAD)

SENSITIVITY
ADJUSTING SCREW

GOVERNOR SPRING

SPEED ADJUSTING NUT

DECREASE SPEED   INCREASE SPEED

Governor assembly

tension. Proper adjustment of the carburetor must be included as part of the governor adjustment procedure. Adjust the carburetor before making any adjustment to the governor linkage.

Adjustment

When the engine is stopped, the carburetor throttle should be held wide open by the tension of the governor spring. In this position, the throttle shaft lever should be just touching the body of the carburetor or clearing it by no more than $\frac{1}{32}$ in. This specification can be obtained by turning the ball joint threads of the connecting linkage and increasing or decreasing the length of the link as necessary.

After making the above check and (if necessary) adjustment, make the engine rpm and governor sensitivity adjustment in the following manner:

Start and run the engine until it reaches normal operating temperatures. Using a tachometer, adjust the speed adjustment nut so that the rpm reading is about 50–100 rpm higher than the desired speed with an average or full-load connected to the engine. Apply the full-load to the engine and check the speed. If the engine begins to hunt, the sensitivity of the governor must be adjusted. The correct sensitivity adjustment will regulate the throttle without causing the engine to hunt. If the drop in rpm is too great when going from no-load to a full-load condition, the end of the spring must be moved closer to the governor shaft. Apply various loads to the engine and observe the rpm fluctuation. If the speed regulation is satisfactory, but the engine tends to hunt, the adjustment is too sensitive and the adjustment screw must be turned so that the end of the spring is moved away from the governor shaft.

NOTE: *Any adjustment to the sensitivity will require that the speed adjustment screw be readjusted. On AK series generators, use a voltmeter connected across the output terminals of the generator in place of the tachometer. Adjust the engine speed to obtain the correct voltage output.*

Recoil Starter

Disassembly

1. Remove the starter assembly from the engine.

2. Turn the starter assembly upside down and remove the retainer ring in the center

with a screwdriver. Hold the washer underneath the retainer so that the spring under the washer will not pop up and be lost.

3. Remove the large washer, the brake spring, the washers under the brake spring, and the friction shoe assembly.

4. To prevent the recoil spring from rotating under the rope sheave, hold the starter cord while removing the four screws that hold the mounting ring and middle flange in place. Remove these two parts. The tension of the recoil spring may now be released slowly by releasing the rope slowly.

5. To prevent the spring from escaping the cover and causing possible injury, lift the rope sheave only $\frac{1}{4}$ in. from the cover and unhook the inside spring loop from the rotor. If the spring does escape, it can be replaced.

6. Clean all of the starter parts in a suitable solvent.

Recoil Spring Replacement

Remove the spring, starting with the innermost loop, by pulling out one loop at a time and holding back the rest. If the spring is to be reused, it must be wound in the direction of crankshaft rotation, starting from the outside coil of the spring.

If you are installing a new spring, you will notice that spring holders are furnished. They simplify installation of the spring. Place the spring in the proper position with the outermost loop engaged around the pin. When the spring is pressed into the cover cavity, the spring holder is released.

Lubricate the mechanism with light grease or, if the engine is used in extremely dusty areas, use powdered graphite on the spring. Avoid all lubrication of the brake washers.

Assembly

1. Assemble the recoil spring and the rope to the sheave.

2. Place the sheave into the cover and hook the inside loop of the spring to the sheave with the help of a screwdriver. Keep a slight tension on the spring to prevent it

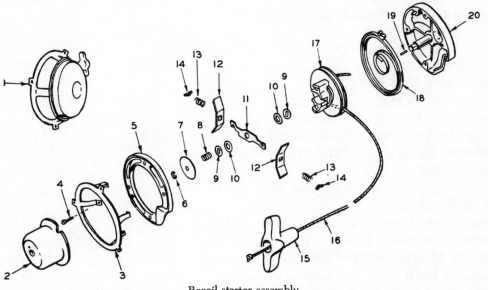

Recoil starter assembly

1. Assembled starter mechanism	11.⎫
2. Cup	12.⎪ Friction shoe assembly
3. Mounting ring	13.⎬
4. Screws	14.⎭
5. Middle flange	15. Rope pull handle
6. Retainer ring	16. Rope
7. Large washer	17. Sheave
8. Spring	18. Recoil spring
9. Washer	19. Centering pin
10. Washer	20. Cover

from becoming unhooked until the middle flange is installed.

3. Install the washers that go under the friction shoes, the friction shoe assembly, the top washers, the spring, the large washer, and the retainer ring.

4. Wind the cord in the proper direction onto the sheave, then add two more turns of the sheave for pre-tension. If the spring is fatigued, it may be necessary to add a few more turns of the sheave to achieve the correct pre-tension of the recoil spring.

5. With tension held on the cord, place the middle flange against the cover and install the mounting ring in position and pointed in the desired direction of pull.

6. Place the starter against the engine blower housing; ensure that the centering pin engages the center hole of the cup and flywheel mounting capscrew while the starter mounting holes align. Pull the pin out farther with a pair of pliers if it fails to engage the center hole of the capscrew. Mount the starter to the blower housing.

	MINIMUM	MAXIMUM
Tappet, Intake Valve - (Cold)	.010"	.012"
Tappet, Exhaust Valve - (Cold)	.010"	.012"
Valve Face, Angle	44°	
Valve Seat, Angle	45°	
Valve Interference Angle	1°	
Valve Stem in Guide - Intake	.0010"	.0025"
Valve Stem in Guide - Exhaust	.0025"	.0040"
Crankshaft End Play	.010"	.015"
Crankshaft Main Bearing	.0035"	.005"
Crankshaft Main Bearing Journal - Standard Size	1.6860"	1.6865"
Crankshaft Rod Bearing Journal - Standard Size	1.3745"	1.3750"
Valve Seat Interference Width	1/32"	3/64"
Camshaft Bearing	.0015"	.0030"
Connecting Rod Bearing	.0015"	.0025"
Piston Pin in Rod - 72°F.	Thumb Push Fit	
Piston Pin in Piston - 72°F.	Hand Push Fit	
Piston to Cylinder - Cast Iron Block .	.0025"	.0045"
Cylinder Bore - Standard Size - Cast Iron Block	2.7505"	2.7515"
Piston Ring Gap - Compression - Cast Iron Block	.007"	.017"
Piston Ring Gap - Oil - Cast Iron Block.	.007"	.017"
Magneto Breaker Point Gap	.022"	
Battery Ignition Point Gap	.016"	.024"
Magneto Pole Shoe Air Gap	.010"	.015"
Spark Plug Gap - Gasoline Fuel	.025"	
Spark Plug Gap - Gas Fuel	.018"	
Ignition Timing Advance - above 1800 rpm.	25° B.T.C.	
- 1800 rpm and slower	19° B.T.C.	
Cylinder Head Screw, Torque	25-30 lb. ft.	
Connecting Rod Screw, Torque	12-15 lb. ft.	

Specifications and torque readings for the AJ series engines

	Minimum	Maximum
Tappet - Intake Valve (at 70° F)	.010''	.012''
Tappet - Exhaust Valve (at 70° F)	.010''	.012''
Valve Face, Angle	.44°	
Valve Seat, Angle	.45°	
Valve Interference Angle	1°	
Valve Stem in Guide - Intake	.0010''	.0025''
Valve Stem in Guide - Exhaust	.0025''	.0040''
Crankshaft End Play	.008''	.012''
Crankshaft Main Bearing	.0030''	.0040''
Crankshaft Main Bearing Journal -		
Standard Size	1.6857'	1.6865'
Valve Seat Width	1/32''	3/64''
Camshaft Bearing	.0015''	.0030''
Connecting Rod Bearing	.0015''	.0025''
Crankshaft Rod Journal - Std Size	1.3742''	1.3750''
Piston Pin in Rod - 72°F	Thumb Push Fit	
Piston Pin in Piston - 72°F	Hand Push Fit	
Piston to Cylinder - Measured at bottom of skirt - 90° from pin	.004''	.005''
Cylinder Bore - Standard Size	2.502''	2.503''
Piston Ring Gap	.006''	.018''
Magneto Breaker Point Gap (full		
separation)	.022''	
Anti-Flicker Breaker Point Gap (full separation)	.020''	
Magneto Pole Shoe Air Gap	.010''	.015''
Spark Plug Gap - Gasoline Fuel	.025''	
Spark Plug Gap - Gas Fuel	.018''	
Ignition Timing Advance - 3000 & 3600 rpm	25°B.T.C.	
1800 rpm Plants	19°B.T.C.	
Cylinder Head Screw, Torque	24-26 lb. ft.	
Connecting Rod Screw, Torque	10-12 lb. ft.	
Oil Base	25-30 lb. ft.	
Timing Gear Cover	15-20 lb. ft.	
Flywheel to Crankshaft	35-40 lb. ft.	
Spark Plug	25-30 lb. ft.	

Specifications and torque readings for the AK series engines

9 . Wisconsin

This chapter covers Wisconsin S-10D, S-12D, and S-14D model engines. These engines are very similar in design, the only difference being the size of the cylinder bore and related parts. All of the engines discussed in this chapter are 4 stroke, horizontal crankshaft engines that are splash lubricated. Crankshaft rotation when viewed from the flywheel side of the engine is clockwise. All engines built after serial number 4225490 are equipped with a camshaft incorporated compression release.

Engine Identification

There is a Wisconsin name plate attached to the blower housing of the engine on which is stamped the model number, serial number, and specification number along with the size and rpm rating. The model, serial, and specification number must be given when obtaining replacement parts for any of the engines. Make certain that the identification plate remains with the engine on which it was originally installed.

Carburetion

A Zenith Model 1408 carburetor is used on these engines. This carburetor is a horizontal side draft, float type carburetor with an adjustable main jet, adjustable idle needle, and a throttle stop screw.

Adjustment

MAIN JET AND IDLE MIXTURE

1. Turn the main jet adjusting screw in until it lightly seats the needle. Do not tighten the screw, as it is very easy to damage the needle by pressing it too hard into the seat.

2. Turn the main jet adjusting screw out 2¼ turns. This is an initial adjustment that enables the engine to be started.

3. Start the engine and run it until normal operating temperature is reached. Adjust the needle with the engine running at normal operating speed until the best performance is obtained.

NOTE: *In cold weather the needle may be opened slightly more to facilitate starting. After the engine is started, return the needle adjustment to its original setting.*

Wisconsin name plate

4. Adjust the idle mixture with the throttle lever in the closed position. The normal setting for the idle mixture adjustment screw is about 1½ turns off its seat. However, it may be necessary to make a slight adjustment to obtain the smoothest idle speed.

5. Adjust the throttle stop screw to obtain the desired idle speed. This adjustment is made at the factory and no further adjustment should be required.

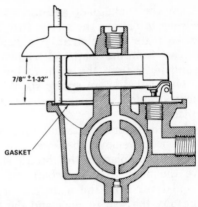

Float level adjustment on the Zenith Model 1408 carburetor

FLOAT LEVEL

1. The float level is adjusted with the float bowl removed and the throttle body held in an inverted position.

2. With the main inlet needle in position and seated, support the float so that the tab or float lever just touches the fuel inlet needle valve.

3. Measure the distance from the bottom of the float, which is in the upper position when the assembly is inverted. The measurement should be ⅞ in. plus or minus ¹⁄₃₂ in.

4. Adjust the float level by removing the float and bending the tab or lever that touches the needle valve with a pair of pliers until the correct level is obtained.

REBUILDING

Disassembly

1. Remove the carburetor from the engine.

2. Invert the carburetor and remove the main jet adjustment assembly, the washer, and the fuel bowl (float bowl). Disassemble the main adjustment needle assembly.

3. Press out the float axle and remove the float.

4. The fuel inlet valve assembly will now fall from the inlet valve seat.

5. Remove the idle adjusting needle and the throttle stop screw, together with each spring, by unscrewing them.

6. Use a screwdriver to remove the fuel inlet valve seat and washer from the fuel inlet opening.

7. Remove the gasket that is installed between the float bowl and the main body of the carburetor.

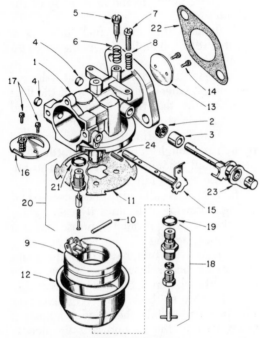

Exploded view of the Zenith Model 1408 carburetor

1. Main body (throttle body)
2. Throttle shaft seal
3. Throttle shaft seal retainer
4. Cup plugs
5. Idle adjustment needle
6. Idle adjustment needle spring
7. Throttle stop screw
8. Throttle stop screw spring
9. Float and hinge assembly
10. Float axle
11. Float bowl-to-body gasket
12. Float/fuel bowl
13. Throttle plate
14. Throttle plate screw
15. Choke lever and shaft assembly
16. Choke plate
17. Choke plate screws
18. Main jet and adjustment assembly
19. Main jet adjustment washer
20. Fuel inlet valve and seat assembly
21. Fuel inlet valve seat gasket
22. Mounting flange gasket
23. Throttle shaft and lever assembly
24. Choke lever friction spring

8. Close the choke plate and use a small screwdriver to remove the screws that hold the choke plate to the shaft. Remove the choke plate from the shaft and remove the shaft and lever from the main carburetor body. The choke detent spring should not be removed unless it is damaged and must be replaced.

9. Close the throttle plate and use a small screwdriver to remove the screws which hold the throttle plate to the throttle shaft; remove the plate from the shaft and the shaft and lever from the main carburetor housing. Pry the seal retainer and the seal off of the shaft hole boss. The shaft hole plugs should not be removed unless they are damaged and must be replaced.

Cleaning and Inspection

1. Clean all of the metal parts in a suitable solvent, removing all carbon deposits from the throttle bore and idle discharge passages. To ensure that all dirt is removed, blow compressed air through all passages in the throttle body and fuel bowl in the reverse direction from normal flow.

NOTE: *Never use wire or a drill to clean jet orifices or idle port openings.*

2. Check the float and make sure that it is not soaked with gasoline. Also check for wear on the float hinge and where the float contacts the inlet needle. Replace the float if any of the above conditions exist.

3. Inspect the main jet adjustment needle and the idle adjusting needle tapered ends to make sure that they are smooth and not grooved from being seated too hard. If there is a groove around the end of the taper, or if it is pitted, replace the needle.

4. Check the fuel inlet valve and seat for wear or damage. Replace the entire assembly as a unit if it doesn't look like new.

5. All gaskets, seals, retainers, and rubber O-rings must be replaced every time the carburetor is overhauled, with the possible exception of the rubber O-rings which can be retained if they are in good condition.

Assembly

1. Assemble the seal and seal retainer to the throttle shaft and insert the throttle shaft into the throttle shaft hole on the engine side of the throttle bore. Press the seal and retainer firmly into place.

2. Position the throttle shaft so that the throttle plate can be slid into the slot and the retaining screws installed and tightened.

3. Install the choke shaft into the choke shaft hole and seat the shaft into the hole in the opposite side. Install the choke plate and tighten the retaining screws.

NOTE: *Be certain that the poppet valve is facing out.*

4. Install the gasket that seals the fuel bowl (float bowl) to the main body of the carburetor.

5. Install the main fuel inlet seat. Tighten it to 100 in. lbs. Insert the valve assembly.

6. Install the float to the float axle and float hinge. Check the operation of the float to make sure that there is no binding and that it moves straight up and down.

7. Install the throttle stop screw and tighten it enough to open the throttle plate just a small amount, but not enough to uncover the second idle discharge port.

8. Install the idle adjustment needle and spring, screwing it in until it seats *lightly* and then backing it out 1½ turns. This is a preliminary setting.

9. Place the fuel bowl onto the main carburetor body. Install the washer onto the fuel bowl. Screw the main jet assembly into the throttle body boss. Tighten it to 100 in. lbs. Install the needle adjusting screw part of the jet and seat the needle lightly. Then back the needle out 2¼ turns. This is a preliminary high speed adjustment.

10. Assemble the carburetor to the engine, using a new carburetor mounting flange gasket if necessary.

Ignition Systems

Wisconsin engines are equipped with either a magneto type ignition, a conventional battery type ignition, or a solid state breakerless ignition system. The magneto and battery systems use the same type of breaker points. The adjustment and replacement procedures are the same for both systems.

BATTERY AND MAGNETO SYSTEMS

Breaker Points Replacement and Adjustment

The breaker points are located in a breaker point box on the side of the engine. The points are actuated by a push pin which is moved by a striker plate on the camshaft. Whenever the points are replaced, be sure to check the condition of the push pin for wear and replace it if necessary.

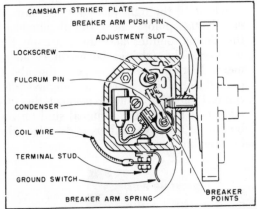

Breaker points and condenser assembly in the breaker box

If oil is present in the breaker box when the cover is removed, this would indicate that the crankcase breather assembly is defective and needs attention. The breather is located on the valve chamber inspection cover. After removing the cover, inspect the condition of the points. If the contacts are pitted or excessively worn, replace the assembly. To remove the breaker assembly, unscrew the fulcrum pin, the adjusting lock screw, and the screw that holds down the clamp for the condenser. Also unhook the breaker assembly from the lead which connects the assembly to the coil.

To replace the assembly:

1. Turn the crankshaft until the breaker push pin is at its lowest position on the camshaft striker plate.

2. Mount the breaker assembly to the engine, sliding the assembly onto the brass fulcrum pin. Be certain that the spring arm is mounted squarely. Tighten the mounting pin and the lock screw with its washer; position the condenser and mount the condenser clamp and screw. Connect the condenser wire into position.

The breaker point gap is to be set at 0.023 in. at the point of maximum opening. However, since the ignition timing is controlled by the setting of the breaker point gap, and the fixed running spark advance is 18° BTDC, it may be necessary that the gap vary slightly from 0.023 in. to gain an accurate spark advance setting.

Ignition Timing

1. Turn the crankshaft until the breaker arm push pin opens the point contacts to the maximum gap.

2. Loosen the lock screw just enough so that the contact support plate can be moved.

3. Position a feeler gauge between the contacts and insert a screwdriver in the adjusting slot. Open or close the points with the screwdriver until there is a slight drag present while sliding the feeler gauge between the point contacts.

4. After the correct gap is obtained, remove the feeler gauge and tighten the lock screw. Recheck the gap to make sure the plate did not move while tightening the lock screw.

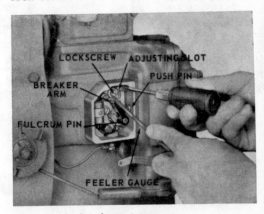

Adjusting the breaker point gap

The above procedure can be used to obtain a fairly accurate ignition timing setting. However, it is recommended that the ignition timing be set by using a self-powered timing light. A more accurate ignition timing setting is possible with the use of a light. Follow the procedure given below for setting the ignition timing with a timing light.

1. Remove the breaker box cover and disconnect the primary coil wire at the bottom of the breaker box.

2. Turn the crankshaft over until the

piston is coming up on its compression stroke. If you remove the spark plug and place your thumb over the hole, you can tell when the piston is coming up on its compression stroke because the air will be forced past your thumb. You can also detect the compression stroke by observing the action of the breaker arm push pin. When the pin starts to come out to open the points, the piston is on the compression stroke.

3. With the piston on the compression stroke, line up the timing pointer and the mark on the flywheel. There is a hole in the back of the flywheel shroud, just to the left of the breaker box, through which the pointer and flywheel mark can be observed.

4. Connect one lead of the timing light to a ground on the engine and the other lead to the terminal stud at the bottom of the breaker box. If the breaker points are closed, the timing light will come on.

5. Loosen the lock screw of the breaker point backing plate just enough so that the backing plate can be moved with some resistance.

6. Place a screwdriver into the adjusting slot and close the points until the timing light comes on.

7. Turn the screwdriver slowly in the opposite direction in the adjusting slot until the light just goes out. Keep the points in this position and tighten the lock screw.

8. Turn the flywheel until the light comes on and continue to turn the flywheel slowly until the light just goes out; stop at exactly this point. Check the alignment of the timing pointer and the mark on the flywheel. They should be in direct alignment. If they are not, repeat the above procedure. If the marks are aligned, disconnect the timing light, connect the primary coil lead to the terminal stud on the bottom of the breaker box, and replace the breaker box cover.

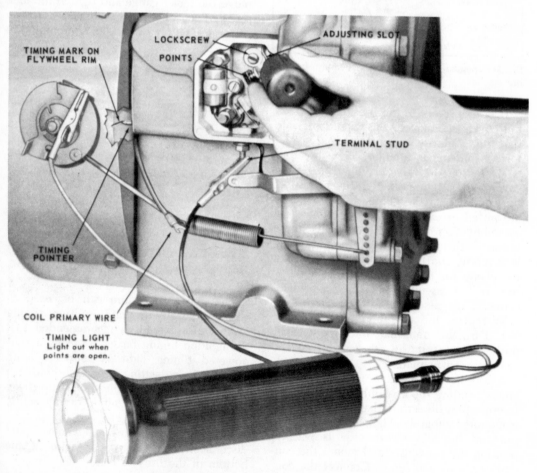

Setting the ignition timing with a continuity light

BREAKERLESS, SOLID STATE IGNITION SYSTEM

The solid state ignition system eliminates breaker points by controlling the ignition spark through the use of a solid state silicon controlled rectifier switch (SCR). Thus the ignition spark is controlled electronically and not mechanically. There are no moving parts in the system except for the magnet ring which is a part of the flywheel. The three basic components of the system are the magnet ring, the stator assembly, and the ignition coil.

The ignition timing is preset and can only be checked for accuracy. It cannot be adjusted.

The operation of the breakerless solid state ignition system is as follows: The magnet ring on the flywheel passes over the coil poles on the stator and generates an alternating current. The alternating current is passed through a diode rectifier which is designed to allow current to pass only in one direction, thus changing the alternating current to direct current. The direct current then passes on to a capacitor which momentarily stores the electrical energy. As the flywheel continues to rotate, the magnets pass over a trigger coil. This generates a small amount of current that is routed to the SCR. The SCR is triggered by this current and allows the current stored in the capacitor to be released and flow on to the primary windings of the ignition coil where the high voltage needed to fire the spark plug is induced into the secondary windings and passed onto the spark plug. The timing of the ignition spark is permanently set by the positioning of the trigger coil in relation to the keyway of the flywheel.

As previously mentioned, the ignition timing can only be checked for accuracy, and not adjusted. The timing can be checked with a neon timing light and a 12 volt battery. Connect the leads of the timing light to the positive and negative terminals of the battery and the spark plug terminal. The ignition spark is retarded to about 10° or 12° BTDC for starting. When the engine starts, the spark automatically advances as the engine speeds up. The spark advance is 20° while the engine is running at 2500 rpm and over.

The timing mark on the flywheel and the positioning of the pointer on the housing of S-10D, S-12D, and S-14D engines is set up for the magneto and battery type igni-

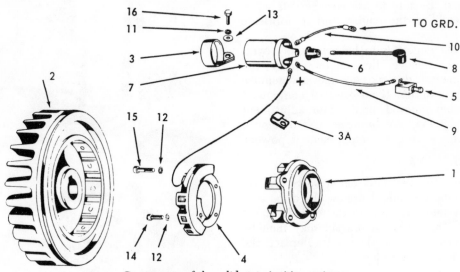

Components of the solid state ignition system

1. Bearing plate assembly
2. Flywheel with magnet ring
3. Strap for the ignition coil
4. Stator assembly
5. Ignition switch
6. Rubber boot for high tension spark plug wire
7. Ignition coil
8. High tension spark plug wire
9. Wire
10. Wire
11. Lockwasher
12. Lockwasher
13. Plain washer
14. Screw
15. Screw
16. Screw

tion systems and represents an 18° spark advance. However, the ignition timing can be checked on engines equipped with the solid state ignition system by using these marks as a reference. When the engine is running at 1000 rpm, the two marks should be in alignment when the timing light illuminates them. To check for the full 20° spark advance at 2500 rpm, speed up the engine to 2500 rpm and aim the timing light at the marks. The mark on the fly-wheel will be illuminated when it is about ⅛ in. above the pointer.

Engine Mechanical

The disassembly, inspection, and assembly of each component of the engine is discussed separately, because many times it is not necessary to disassemble the entire engine. The order in which the disassembly procedures are given may be changed to suit the job.

Whenever the engine is either partially or completely disassembled, all of the parts removed should be thoroughly cleaned. Be sure to use new gaskets when reassembling the engine and to lubricate all bearings.

If the engine is to be completely overhauled, remove the engine from the machinery it drives or operates and remove any accessories. If an external component is to be removed, or a minor adjustment made, it may not be necessary to remove the engine from the equipment it powers.

FUEL TANK

Close the fuel tank outlet valve and remove the fuel line. Unscrew the three nuts that retain the tank to the cylinder head bolts. The tank and bracket may then be removed as a complete unit. Replace the tank in reverse order.

AIR CLEANER AND CARBURETOR

Unscrew the wing nut and remove the air cleaner. Remove the breather line at the inspection cover, the throttle rod clip at the governor lever, and the fuel line. Unscrew the two nuts which hold the carburetor to the engine and remove the car-

buretor and air cleaner bracket as one. Replace in reverse order.

STARTER SHEAVE AND FLYWHEEL SHROUD

Remove the starter sheave by removing the three screws and washers which retain it to the flywheel. Remove the top cover and the cylinder side shroud. Disconnect the governor spring and remove the four screws that hold the flywheel shroud to the back plate. The entire flywheel shroud may now be removed. The back plate can be removed, if necessary, only after the flywheel is removed. Reassemble in the reverse order.

CYLINDER HEAD

Remove the spark plug and unscrew the five cap screws and three studs that attach the cylinder head. Remove the cylinder head and gasket. Clean the carbon from the combustion chamber and all dirt from the cooling fins. When installing the cylinder head, please note that the cylinder head gasket is not symmetrical. The larger inside radius around the valves is to be

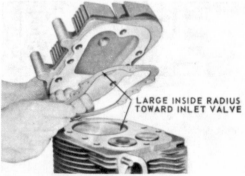

LARGE INSIDE RADIUS
TOWARD INLET VALVE

Removal and installation of the cylinder head and gasket

placed at the inlet valve. Apply a mixture of oil and graphite to the threads of the cap screws and studs before screwing them into the head. Tighten the studs and screws in three alternate stages and in a criss-cross pattern to 16 ft lbs, 24 ft lbs, and finally to 32 ft lbs.

VALVES AND VALVE SEAT INSERTS

Remove the valve inspection cover which is also the breather assembly. Use a valve spring compressor to compress the valve springs. Remove the valve spring retainers, the compressor, and the valve springs; take the valve out from the top of the cylinder block. Clean all carbon deposits from the valves, seats, ports, and guides. Inspect the condition of the valves, stems, guides, and seat, looking for burned, pitted, scored, or warped surfaces.

Removal and installation of the valves, retainers, and springs

Removal and installation of the valve seat inserts using special tools

The exhaust valve and seat are made of stellite. A valve rotator is used on the exhaust valve only. Clean the valve rotator and make sure that it operates properly.

Both valve seats are removable and replaceable. Valve seats are removed by means of a special puller. After the new seats are installed, they should be ground to the proper angle.

Before grinding the seats or valves, check the valve-to-guide clearance. If it is excessive, the guides can be pressed out and new ones pressed into place. Special tools are required to replace valve guides. Try replacing the valve to gain the proper stem-to-guide clearance before installing new guides.

After grinding the valves, they must be lapped into place until there is a solid ring around the face of the valve. Clean the valve and wash the cylinder block with hot water and soap. Next wipe down the cylinder walls with clean light engine oil. This should be performed especially if the cylinder bore has been honed.

FLYWHEEL

If the flywheel is to be removed, loosen the retaining nut before the gear cover on the opposite end is removed.

NOTE: *Do not try to loosen the flywheel after the gear cover is removed. Do not strike the crankshaft when it is not supported by the gear cover.*

To remove the flywheel, first straighten the tab of the washer under the flywheel retaining nut. Place the correct size wrench on the flywheel retaining nut and strike the wrench sharply with a hammer to loosen the nut. Do not remove the nut completely, just unscrew it until it is flush with the end of the crankshaft. Grasp the flywheel and pull it outward and at the same time strike the end of the crankshaft with a soft hammer. This will loosen the flywheel from the tapered end of the crankshaft. Loosen the flywheel, but do not remove it at this point. It is necessary for the flywheel to remain on the crankshaft and support it while the gear cover and connecting rod are removed. Remove the flywheel only after the piston and connecting rod are removed.

When reassembling the engine, install the flywheel after the crankshaft is installed. Make sure that the woodruff key is in place before positioning the flywheel onto the crankshaft. Do not drive the flywheel onto the crankshaft by striking it with a hammer. Place a small length of pipe against the hub of the flywheel and tap the end of the pipe with a soft hammer until the flywheel is seated on the crankshaft taper. Assemble the washer and nut to the crankshaft with the tab of the washer inserted into the keyway of the flywheel. Tighten the nut only enough to hold the flywheel in place. Only after the crankshaft endplay has been adjusted is the flywheel nut to be tightened by sharply striking the wrench with a soft hammer. Bend the tab of the washer up against the nut.

GEAR COVER

To remove the gear cover, unscrew the gear cap screws and remove the governor lever. Tap the two dowel pins lightly from the crankcase side to break the cover loose from the crankcase.

NOTE: *A steel ball for the end thrust of the camshaft will most likely fall out when the cover is removed. Remove the spring from the end of the camshaft so it won't be lost.*

Removing the gear cover

To reassemble the gear cover to the engine, position the spring into the end of the camshaft and lubricate the bearings, gears and tappets. Tap the dowel pins into the crankcase until they protrude about ⅛ in. from the mounting flange face. Place a finger full of grease into the hole in the cover to retain the camshaft spring and ball in place. Lubricate the lip of the oil seal with engine oil. Assemble the gear

cover to the engine by tapping it into place on the dowel pins with a soft hammer. Make sure that the timing marks on the camshaft and crankshaft stay aligned when the end of the camshaft is pressed into the bearing hole in the gear cover. Install the governor yoke so that it straddles the governor shaft extension and bears against the thrust sleeve. Install the gear cover retaining screws and tighten them to 18 ft lbs of torque.

GOVERNOR FLYWEIGHT MECHANISM

The governor gear and flyweight assembly rotates on a stationary pin which is pressed into the crankcase and held in place by a snap ring.

To remove the assembly, first spread the flyweights apart and remove the governor thrust sleeve. Remove the snap ring with a pair of snap ring pliers. The governor gear and flyweights will now slip off the shaft.

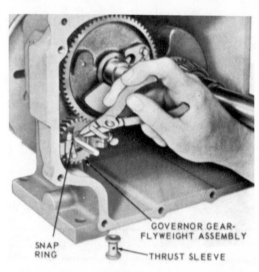

Removing the snap-ring from the governor shaft

Reassemble the governor in the reverse order of removal. There should be a clearance of 0.003–0.005 in. between the gear hub and the face of the governor shaft boss in the crankcase. This clearance is adjusted by tapping the governor shaft in the required direction. There should be a clearance of 0.0005–0.002 in. between governor shaft and the gear. When the clearance is 0.005 in. or less, the worn parts should be replaced.

CONNECTING ROD, PISTON AND PISTON RINGS

Unscrew the two cap bolts which hold the connecting rod cap to the connecting rod. The oil dipper will come off with the cap screws. Tap the ends of the bolts to loosen the connecting rod cap.

Remove all deposits from the cylinder that might hinder the removal of the piston. This is done with a ridge reamer.

Removing the piston and connecting rod assembly

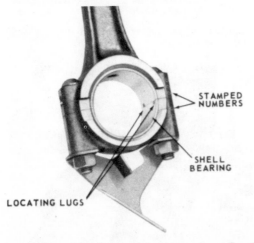

LOCATING LUGS

STAMPED NUMBERS

SHELL BEARING

Assembly and alignment of the connecting rod and bearing

Turn the crankshaft until the piston is at the top of the cylinder and push the connecting rod and piston assembly up and out of the engine.

The connecting rod bearing is a shell type and is removable. Be sure to assemble the bearing to the connecting rod so that the locating lug on both bearing halves are on the same side.

The piston skirt is elliptical in shape. When measuring the piston-to-cylinder wall clearance, you must take the measurement at the bottom of the piston skirt thrust face. The thrust faces of the piston skirt are located at a 90° angle from the piston pin hole axis.

Install the piston rings so that the ring gaps are 90° apart around the circumference of the piston. A ring expander tool should be used to remove and install piston rings. If the tool is not available, the rings can be installed by placing the open end of the ring into the appropriate groove and working the ring down over the piston. Install the bottom oil control ring first, the scraper ring second and the compression ring last. Be careful not to bend or distort the rings in any way. A notch mark or the word "top" will be stamped on each ring so as to identify which side of the ring should

PISTON TO CYLINDER AT PISTON SKIRT THRUST FACES	MODELS S-10D, S-12D .0025 to .003"	MODEL S-14D .0025 to .004"
PISTON RING GAP		.010 to .020"
PISTON RING SIDE CLEARANCE IN GROOVES	TOP RING	.002 to .004"
	2nd RING	.002 to .004"
	OIL RING	.0015 to .0035"
Connecting Rod to Crank Pin — Side Clearance		.004 to .013"
Connecting Rod Shell Bearing to Crank Pin		.0005 to .0015"
PISTON PIN TO CONNECTING ROD BUSHING		.0005 to .0011"
PISTON PIN TO PISTON		.0000 to .0008" tight

$\frac{5}{64}$ R.   1.4990 / 1.4984 DIA.

1.255 / 1.250   WIDTH

STANDARD CRANK PIN DIMENSIONS

Rebuilding specifications

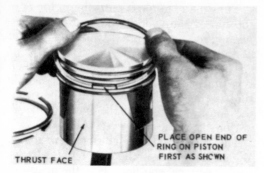

Assembling a piston ring to the piston

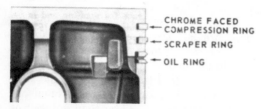

Positioning of the piston rings

face the top of the piston. Before installing the piston assembly into the cylinder, oil the rings, cylinder wall, rod bearings, wrist pin and the piston itself. Use a ring compressor to install the piston assembly into the cylinder bore.

The piston/connecting rod assembly is mounted on the crankshaft with the stamped number on the connecting rod facing toward the open end of the crankcase. Install the conecting rod bearing cap in the same manner, with the stamped numbers facing toward the open end of the crankcase. Rotate the crankshaft until it is at the bottom of its stroke. Tap the piston down so that the connecting rod seats onto the crankpin. Install the oil dipper so that the cap screws are accessible from the open end of the crankcase. Tighten the cap screws to 22 ft lbs.

CYLINDER BLOCK

The cylinder block is a replaceable item. It should not be removed from the crankcase unless the cylinder bore is worn more than 0.005 in. oversize or damaged in some way (cracked, broken cooling fins, etc.). Oversize pistons and rings are available in 0.010 in., 0.020 in., and 0.030 in. sizes. When the cylinder block is reinstalled on the crankcase, tighten the mounting nuts to 40–50 ft lbs. The cap screw in the valve

spring box should be tightened to only 32 ft lbs.

CAMSHAFT AND VALVE TAPPETS

When removing the camshaft, turn the engine over on its side and push the tappets away from the camshaft so that they will clear the camshaft lobes when the camshaft is removed. The valves must be removed for this operation. After the camshaft is removed, remove the valve tappets and inspect them for wear. The body of the tappet must be 0.6245–0.6235 in. in diameter. When the tappet is installed in the guide hole, there should be 0.005–0.0025 in. clearance.

The valve tappets must be installed before the camshaft. When the camshaft is mounted, align the marks on the camshaft with the marked gear tooth on the crankshaft gear. Place the camshaft thrust spring in the end of the camshaft before installing the gear cover.

The valve tappet clearance is adjusted

Alignment of the timing marks on the crankshaft and camshaft gears

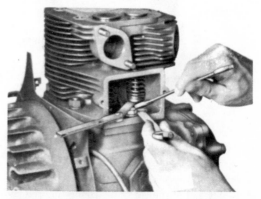

Adjusting valve tappet clearance

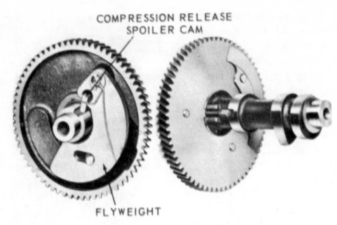

COMPRESSION RELEASE
SPOILER CAM

FLYWEIGHT

Compression release mechanism

with the engine cold and with the tappets in their lowest position. The inlet valve clearance must be 0.007 in. and the exhaust valve clearance must be 0.016 in.

NOTE: *Make sure that the exhaust tappet is not positioned on the compression release mechanism.*

BREAKER PUSH PIN AND BUSHING

The breaker push pin should be replaced if there is an indication of excessive wear. A repair kit is available and includes a new pin, bushing, and spring along with instructions. Mount the spring under the head of the push pin and insert the pin into the guide hole with the rounded end of the pin toward the camshaft striker plate.

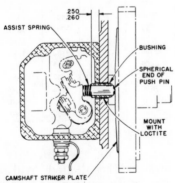

ASSIST SPRING

.250
.260

BUSHING

SPHERICAL
END OF
PUSH PIN

MOUNT
WITH
LOCTITE

CAMSHAFT STRIKER PLATE

Breaker point push pin and bushing

CRANKSHAFT

The crankshaft is removed after the gear cover has been removed, the connecting rod disconnected and raised up out of the way. Remove the flywheel nut, flywheel, and the woodruff key. The crankshaft may now be

pulled out of the open end of the crankcase. When reinstalling the crankshaft, mount the flywheel after the crankshaft is inserted into the crankcase. The flywheel supports the crankshaft while the connecting rod is attached. The flywheel nut is tightened only after the gear cover is installed and before the end-play is adjusted.

STATOR PLATE AND END-PLAY ADJUSTMENT

The end-play of the crankshaft is adjusted by the application of various size gaskets behind the stator plate, which doubles in function as the front bearing support and an adaptor for the magneto coil. The stator plate should not be removed from the crankcase unless it has to be replaced.

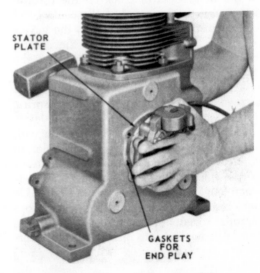

STATOR
PLATE

GASKETS
FOR
END PLAY

Removal and installation of the stator plate

To remove the stator plate, remove the four retaining screws and tap the plate from the inside until it falls off. Reassemble the stator plate to the crankcase using new gaskets with the same total thickness as those originally installed. The stator plate capscrews are to be tightened to 18 ft lbs.

Check the crankshaft end-play after the gear cover and flywheel are mounted. The crankshaft end-play should be between 0.001 in. and 0.004 in. and is to be measured only when the engine is cold. Crankshaft end-play is measured with a dial indicator mounted on the PTO side of the crankshaft and a lever prying behind the flywheel. If new crankshaft roller bearings have been installed, they must be properly seated by tapping the ends of the crankshaft with a lead hammer before measuring the crankshaft end-play.

GOVERNOR

Adjustment

Adjustment of the governor is made when the engine is not operating. Disconnect the control rod from the governor lever and push the rod toward the carburetor as far as it will go. This puts the carburetor throttle lever in a wide open postion. Next, move the governor lever in the same manner, so that the end in which the governor rod fastens is moved up toward the carburetor as far as it will go. Holding both of these parts in the position mentioned above, screw the control rod in or out of the swivel block on the throttle lever until the bent end of the control rod can be inserted into the hole in the governor lever. Be sure to replace the control rod clip and tighten the lock nut on the swivel block after making this adjustment.

The governor controlled engine speed is regulated by hooking the governor spring in the various holes of the governor lever, and then adjusting the spring tension with an adjusting screw. There are 7 holes in the governor lever, with #1 hole closest to the fulcrum shaft. There are two different length adjusting screws available which allow the full range of engine operating speeds to be attained. After the spring is hooked into one of the holes cor-

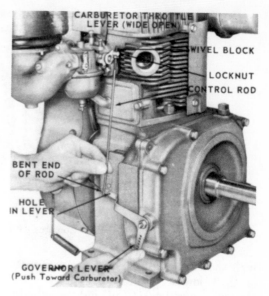

Adjustment of the governor mechanism

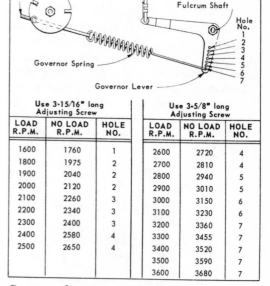

Use 3-15/16" long Adjusting Screw			Use 3-5/8" long Adjusting Screw		
LOAD R.P.M.	NO LOAD R.P.M.	HOLE NO.	LOAD R.P.M.	NO LOAD R.P.M.	HOLE NO.
1600	1760	1	2600	2720	4
1800	1975	2	2700	2810	4
1900	2040	2	2800	2940	5
2000	2120	2	2900	3010	5
2100	2260	3	3000	3150	6
2200	2340	3	3100	3230	6
2300	2400	3	3200	3360	7
2400	2580	4	3300	3455	7
2500	2650	4	3400	3520	7
			3500	3590	7
			3600	3680	7

Governor adjustment specifications

responding to the desired load engine speed, the corresponding no-load engine speed is then attained by regulating the spring tension with the adjusting screw. Each time the adjusting screw is turned, the spring will have to be disconnected from the governor lever.

Specifications for Wisconsin Engines

Model	Standard Cylinder Bore	Piston Skirt-to-Bore Clearance	Crankpin Diameter	Bearing-to-Crankpin Clearance	Valve Tappet Clearance ▲		Running Spark Advance	Maximum Operating Speed
					Inlet	Exhaust		
S-10D	3.5005–3.4995	0.0025–0.0030						
S-12D	3.2505–3.2495	0.0025–0.0030	1.4990–1.4984	0.0005–0.0015①	0.007	0.016	18°②	3600 rpm
S-14D	3.7505–3.7495	0.0025–0.0040						

▲ Engine cold
① Shell bearing-to-crankpin clearance
② 2000 rpm or above

Torque Specifications (ft lbs)

Connecting Rod	Cylinder Block	Cylinder Head	Gear Cover	Stator Plate	Spark Plug
22	50	32	22	18	30

Crankshaft End-Play (cold)	Breaker Point Gap	Piston Pin-to-Piston Clearance	Piston Pin-to-Connecting Rod Clearance	Spark Plug Gap	Valve Seat and Face Angle	Valve-to-Guide Clearance	
						Inlet	Exhaust
0.002–0.004	0.015	0.0000–0.0008	0.0005–0.0010	0.030	45°	0.001–0.003	0.003–0.005

Appendix

General Conversion Table

Multiply by	To convert	To	
2.54	Inches	Centimeters	.3937
30.48	Feet	Centimeters	.0328
.914	Yards	Meters	1.094
1.609	Miles	Kilometers	.621
.645	Square inches	Square cm.	.155
.836	Square yards	Square meters	1.196
16.39	Cubic inches	Cubic cm.	.061
28.3	Cubic feet	Liters	.0353
.4536	Pounds	Kilograms	2.2045
4.546	Gallons	Liters	.264
.068	Lbs./sq. in. (psi)	Atmospheres	14.7
.138	Foot pounds	Kg. m.	7.23
1.014	H.P. (DIN)	H.P. (SAE)	.9861
——	To obtain	From	Multiply by

Note: 1 cm. equals 10 mm.; 1 mm. equals .0394″.

Conversion—Common Fractions to Decimals and Millimeters

INCHES			INCHES			INCHES		
Common Fractions	Decimal Fractions	Millimeters (approx.)	Common Fractions	Decimal Fractions	Millimeters (approx.)	Common Fractions	Decimal Fractions	Millimeters (approx.)
1/128	.008	0.20	11/32	.344	8.73	43/64	.672	17.07
1/64	.016	0.40	23/64	.359	9.13	11/16	.688	17.46
1/32	.031	0.79	3/8	.375	9.53	45/64	.703	17.86
3/64	.047	1.19	25/64	.391	9.92	23/32	.719	18.26
1/16	.063	1.59	13/32	.406	10.32	47/64	.734	18.65
5/64	.078	1.98	27/64	.422	10.72	3/4	.750	19.05
3/32	.094	2.38	7/16	.438	11.11	49/64	.766	19.45
7/64	.109	2.78	29/64	.453	11.51	25/32	.781	19.84
1/8	.125	3.18	15/32	.469	11.91	51/64	.797	20.24
9/64	.141	3.57	31/64	.484	12.30	13/16	.813	20.64
5/32	.156	3.97	1/2	.500	12.70	53/64	.828	21.03
11/64	.172	4.37	33/64	.516	13.10	27/32	.844	21.43
3/16	.188	4.76	17/32	.531	13.49	55/64	.859	21.83
13/64	.203	5.16	35/64	.547	13.89	7/8	.875	22.23
7/32	.219	5.56	9/16	.563	14.29	57/64	.891	22.62
15/64	.234	5.95	37/64	.578	14.68	29/32	.906	23.02
1/4	.250	6.35	19/32	.594	15.08	59/64	.922	23.42
17/64	.266	6.75	39/64	.609	15.48	15/16	.938	23.81
9/32	.281	7.14	5/8	.625	15.88	61/64	.953	24.21
19/64	.297	7.54	41/64	.641	16.27	31/32	.969	24.61
5/16	.313	7.94	21/32	.656	16.67	63/64	.984	25.00
21/64	.328	8.33						

Conversion—Millimeters to Decimal Inches

mm	inches	mm	inches	mm	inches	mm	inches	mm	inches
1	.039 370	31	1.220 470	61	2.401 570	91	3.582 670	210	8.267 700
2	.078 740	32	1.259 840	62	2.440 940	92	3.622 040	220	8.661 400
3	.118 110	33	1.299 210	63	2.480 310	93	3.661 410	230	9.055 100
4	.157 480	34	1.338 580	64	2.519 680	94	3.700 780	240	9.448 800
5	.196 850	35	1.377 949	65	2.559 050	95	3.740 150	250	9.842 500
6	.236 220	36	1.417 319	66	2.598 420	96	3.779 520	260	10.236 200
7	.275 590	37	1.456 689	67	2.637 790	97	3.818 890	270	10.629 900
8	.314 960	38	1.496 050	68	2.677 160	98	3.858 260	280	11.032 600
9	.354 330	39	1.535 430	69	2.716 530	99	3.897 630	290	11.417 300
10	.393 700	40	1.574 800	70	2.755 900	100	3.937 000	300	11.811 000
11	.433 070	41	1.614 170	71	2.795 270	105	4.133 848	310	12.204 700
12	.472 440	42	1.653 540	72	2.834 640	110	4.330 700	320	12.598 400
13	.511 810	43	1.692 910	73	2.874 010	115	4.527 550	330	12.992 100
14	.551 180	44	1.732 280	74	2.913 380	120	4.724 400	340	13.385 800
15	.590 550	45	1.771 650	75	2.952 750	125	4.921 250	350	13.779 500
16	.629 920	46	1.811 020	76	2.992 120	130	5.118 100	360	14.173 200
17	.669 290	47	1.850 390	77	3.031 490	135	5.314 950	370	14.566 900
18	.708 660	48	1.889 760	78	3.070 860	140	5.511 800	380	14.960 600
19	.748 030	49	1.929 130	79	3.110 230	145	5.708 650	390	15.354 300
20	.787 400	50	1.968 500	80	3.149 600	150	5.905 500	400	15.748 000
21	.826 770	51	2.007 870	81	3.188 970	155	6.102 350	500	19.685 000
22	.866 140	52	2.047 240	82	3.228 340	160	6.299 200	600	23.622 000
23	.905 510	53	2.086 610	83	3.267 710	165	6.496 050	700	27.559 000
24	.944 880	54	2.125 980	84	3.307 080	170	6.692 900	800	31.496 000
25	.984 250	55	2.165 350	85	3.346 450	175	6.889 750	900	35.433 000
26	1.023 620	56	2.204 720	86	3.385 820	180	7.086 600	1000	39.370 000
27	1.062 990	57	2.244 090	87	3.425 190	185	7.283 450	2000	78.740 000
28	1.102 360	58	2.283 460	88	3.464 560	190	7.480 300	3000	118.110 000
29	1.141 730	59	2.322 830	89	3.503 903	195	7.677 150	4000	157.480 000
30	1.181 100	60	2.362 200	90	3.543 300	200	7.874 000	5000	196.850 000

To change decimal millimeters to decimal inches, position the decimal point where desired on either side of the millimeter measurement shown and reset the inches decimal by the same number of digits in the same direction. For example, to convert .001 mm into decimal inches, reset the decimal behind the 1 mm (shown on the chart) to .001; change the decimal inch equivalent (.039″ shown) to .00039″).

Tap Drill Sizes

	National Fine or S.A.E.			National Coarse or U.S.S.	
Screw & Tap Size	Threads Per Inch	Use Drill Number	Screw & Tap Size	Threads Per Inch	Use Drill Number
No. 5	44	37	No. 5	40	39
No. 6	40	33	No. 6	32	36
No. 8	36	29	No. 8	32	29
No. 10	32	21	No. 10	24	25
No. 12	28	15	No. 12	24	17
$1/4$	28	3	$1/4$	20	8
$5/16$	24	1	$5/16$	18	F
$3/8$	24	Q	$3/8$	16	$5/16$
$7/16$	20	W	$7/16$	14	U
$1/2$	20	$29/64$	$1/2$	13	$27/64$
$9/16$	18	$33/64$	$9/16$	12	$31/64$
$5/8$	18	$37/64$	$5/8$	11	$17/32$
$3/4$	16	$11/16$	$3/4$	10	$21/32$
$7/8$	14	$13/16$	$7/8$	9	$49/64$
$1 1/8$	12	$1 3/64$	1	8	$7/8$
$1 1/4$	12	$1 11/64$	$1 1/8$	7	$63/64$
$1 1/2$	12	$1 27/64$	$1 1/4$	7	$1 7/64$
			$1 1/2$	6	$1 11/32$

Decimal Equivalent Size of the Number Drills

Drill No.	Decimal Equivalent	Drill No.	Decimal Equivalent	Drill No.	Decimal Equivalent
80	.0135	53	.0595	26	.1470
79	.0145	52	.0635	25	.1495
78	.0160	51	.0670	24	.1520
77	.0180	50	.0700	23	.1540
76	.0200	49	.0730	22	.1570
75	.0210	48	.0760	21	.1590
74	.0225	47	.0785	20	.1610
73	.0240	46	.0810	19	.1660
72	.0250	45	.0820	18	.1695
71	.0260	44	.0860	17	.1730
70	.0280	43	.0890	16	.1770
69	.0292	42	.0935	15	.1800
68	.0310	41	.0960	14	.1820
67	.0320	40	.0980	13	.1850
66	.0330	39	.0995	12	.1890
65	.0350	38	.1015	11	.1910
64	.0360	37	.1040	10	.1935
63	.0370	36	.1065	9	.1960
62	.0380	35	.1100	8	.1990
61	.0390	34	.1110	7	.2010
60	.0400	33	.1130	6	.2040
59	.0410	32	.1160	5	.2055
58	.0420	31	.1200	4	.2090
57	.0430	30	.1285	3	.2130
56	.0465	29	.1360	2	.2210
55	.0520	28	.1405	1	.2280
54	.0550	27	.1440		

Decimal Equivalent Size of the Letter Drills

Letter Drill	Decimal Equivalent	Letter Drill	Decimal Equivalent	Letter Drill	Decimal Equivalent
A	.234	J	.277	S	.348
B	.238	K	.281	T	.358
C	.242	L	.290	U	.368
D	.246	M	.295	V	.377
E	.250	N	.302	W	.386
F	.257	O	.316	X	.397
G	.261	P	.323	Y	.404
H	.266	Q	.332	Z	.413
I	.272	R	.339		

ANTI-FREEZE INFORMATION

Freezing and Boiling Points of Solutions
According to Percentage of Alcohol or Ethylene Glycol

Freezing Point of Solution	Alcohol Volume %	Alcohol Solution Boils at	Ethylene Glycol Volume %	Ethylene Glycol Solution Boils at
20°F.	12	196°F.	16	216°F.
10°F.	20	189°F.	25	218°F.
0°F.	27	184°F.	33	220°F.
−10°F.	32	181°F.	39	222°F.
−20°F.	38	178°F.	44	224°F.
−30°F.	42	176°F.	48	225°F.

Note: above boiling points are at sea level. For every 1,000 feet of altitude, boiling points are approximately 2°F. lower than those shown. For every pound of pressure exerted by the pressure cap, the boiling points are approximately 3°F. higher than those shown.